T0350138

COMPUTATIONAL METHODS WITH APPLICATIONS IN BIOINFORMATICS ANALYSIS

ADVANCED SERIES IN ELECTRICAL AND COMPUTER ENGINEERING

Editor: W. K. Chen *(University of Illinois, Chicago, USA)*

Published:

Vol. 7: Diagnostic Measurements in LSI/VLSI Integrated Circuits Production
by A. Jakubowski, W. Marciniak and H. Przewlocki

Vol. 8: An Introduction to Control Systems (Second Edition)
by K. Warwick

Vol. 9: Orthogonal Functions in Systems and Control
by K. B. Datta and B. M. Mohan

Vol. 10: Introduction to High Power Pulse Technology
by S. T. Pai and Q. Zhang

Vol. 11: Systems and Control: An Introduction to Linear, Sampled and Nonlinear Systems
by T. Dougherty

Vol. 12: Protocol Conformance Testing Using Unique Input/Output Sequences
by X. Sun, C. Feng, Y. Shen and F. Lombardi

Vol. 13: Semiconductor Manufacturing Technology
by C. S. Yoo

Vol. 14: Linear Parameter-Varying System Identification: New Developments and Trends
by Paulo Lopes dos Santos, Teresa Paula Azevedo Perdicoúlis, Carlo Novara, Jose A. Ramos and Daniel E. Rivera

Vol. 15 Active Network Analysis: Feedback Amplifier Theory (Second Edition)
by Wai-Kai Chen (University of Illinois, Chicago, USA)

Vol. 16: Design Techniques for Integrated CMOS Class-D Audio Amplifiers
by Adrian I. Colli-Menchi, Miguel A. Rojas-Gonzalez and Edgar Sanchez-Sinencio

Vol. 18: Broadband Matching: Theory and Implementations (Third Edition)
by Wai-Kai Chen

Vol. 20: Computational Methods with Applications in Bioinformatics Analysis
edited by Jeffrey J. P. Tsai and Ka-Lok Ng

For the complete list of titles in this series, please visit
http://www.worldscientific.com/series/asece

Advanced Series in Electrical and Computer Engineering – Vol. 20

COMPUTATIONAL METHODS WITH APPLICATIONS IN BIOINFORMATICS ANALYSIS

Editors

Jeffrey J. P. Tsai

Ka-Lok Ng

Asia University, Taiwan

World Scientific

NEW JERSEY · LONDON · SINGAPORE · BEIJING · SHANGHAI · HONG KONG · TAIPEI · CHENNAI

Published by

World Scientific Publishing Co. Pte. Ltd.
5 Toh Tuck Link, Singapore 596224
USA office: 27 Warren Street, Suite 401-402, Hackensack, NJ 07601
UK office: 57 Shelton Street, Covent Garden, London WC2H 9HE

Library of Congress Cataloging-in-Publication Data
Names: Tsai, Jeffrey J.-P., editor. | Ng, Ka-Lok, editor.
Title: Computational methods with applications in bioinformatics analysis /
edited by: Jeffrey J.P. Tsai (Asia University, Taiwan), Ka-Lok Ng (Asia University, Taiwan).
Description: New Jersey : World Scientific, 2017. | Series: Advanced series in
electrical and computer engineering ; volume 20
Identifiers: LCCN 2016059286 | ISBN 9789813207974 (hc : alk. paper)
Subjects: LCSH: Bioinformatics--Mathematics. | Computational biology.
Classification: LCC QH324.2 .C6359 2017 | DDC 570.285--dc23
LC record available at https://lccn.loc.gov/2016059286

British Library Cataloguing-in-Publication Data
A catalogue record for this book is available from the British Library.

Desk Editor: Herbert Moses

Typeset by Stallion Press
Email: enquiries@stallionpress.com

Printed in Singapore

Preface

In the past decade, we have seen mounting evidence of the usefulness of computational approaches to bioinformatics and biomedical research. Semantic computing, machine learning classifiers, clustering algorithms and mathematical methods have been particularly influential in contributing insights in the era of post-genomics research. In the post-genomics era, computational biology will certainly play an increasingly important role in elucidating the underlying features and mechanisms in biological systems.

The use of advanced computational approaches is an effective mean to address difficult problems in bioinformatics and biomedical research. An advantage of applying computational methods is to reduce the cost and time compared to doing real experiments; hence, there are needs for developing and advancing such area of research. As the volume of biomedical data increases, so does demand for establishing and building up more computational methods for solving a wide range of applications.

This book is a collection of 10 chapters written by international researchers with expertise in microarray data analysis, semantic computing, dynamics modelling of biomolecular interactions, fuzzy integral, molecular simulation and machine learning algorithms. In particular, a variety of computational methods are presented to different areas of studies in biomedicine and bioinformatics analysis; such as, time series data analysis, ensemble clustering, protein and nucleic acid interaction, host–pathogen interaction system, RNA network in autoimmune diseases, multiple omics data integration, T-cell epitope and cytometry.

In short, this book presents the use of advanced computational methods for a range of biomedical applications. Instead of focusing on problems in molecular biology, sequence analysis or cancer-related research, the book pay attention to a range of biomedical issues rather than the fundamentals.

Jeffrey J.P. Tsai
Ka-Lok Ng
Taichung, 2017

Acknowledgment

We are grateful for all the authors for their efforts and involvement in producing this book. This book would not have been possible without the financial support by the Ministry of Science and Technology of Taiwan (MOST) under contract number MOST 105-2632-E-468 -002, and the support from Asia University. Finally, we would like to thank the editorial and production staffs at World Scientific Publishing, in particular Steven Patt, Herbert Moses and Rajesh Babu, for making this book possible.

About the Authors

Jeffrey J. P. Tsai received a PhD degree in Computer Science from the Northwestern University, Evanston, Illinois. He is currently the President and Chair Professor of Asia University, Taiwan. Dr. Tsai was a Professor of Computer Science and the Director of the Distributed Real-Time Intelligent Systems Laboratory at the University of Illinois, Chicago. He was also an Adjunct Professor at Tulane University, a Visiting Professor at Stanford University, a Visiting Scholar at the University of California at Berkeley, and a Senior Research Fellow of IC2 at the University of Texas at Austin. His research interests include bioinformatics, intrusion detection, knowledge-based software engineering, formal modelling and verification, distributed real-time systems, sensor networks, ubiquitous computing, services computing, and intelligent agents. His research has been supported by US NSF, DARPA, USAF Rome Laboratory, Department of Defense, Army Research Laboratory, Motorola, Fujitsu, and Gtech. The technology on knowledge-based software engineering developed by him and his research team resulted the world's first complete transformation of an embedded software product in 1993 and is now used to produce communication software systems worldwide. Tsai coauthored Knowledge-Based Software Development for Real-Time Distributed Systems (World Scientific, 1993), Distributed Real-Time Systems (Wiley, 1996), Compositional Verification of Concurrent and Real-Time Systems (Springer/Kluwer, 2002), Security Modeling and Analysis of Mobile Agent Systems (Imperial College Press, 2006), Intrusion Detection: A Machine Learning Approach (Imperial College Press, 2010), and coedited Monitoring and Debugging of Distributed Real-Time Systems (IEEE Computer Society Press, 1995), Machine Learning Applications in Software Engineering (WSP, 2005),

Ubiquitous Intelligence and Computing (Springer, 2006), Machine Learning in Cyber Trust: Security, Privacy, Reliability (Springer, 2009). Dr. Tsai was the Conference Co-Chair of the 16th IEEE International Symposium on Software Reliability Engineering, the 9th IEEE International Symposium on Multimedia, the 1st IEEE International Conference on Sensor Networks, Ubiquitous, and Trustworthy Computing, and the 3rd IFIP International Conference on Ubiquitous Intelligence and Computing. From 2000 to 2003, Dr. Tsai chaired the IEEE/CS Technical Committee on Multimedia Computing and served on the steering committee of the IEEE Transactions on Multimedia. From 1994 to 1999, he was an Associate Editor of the IEEE Transactions on Knowledge and Data Engineering and he is currently an Associate Editor of the IEEE Transactions on Services Computing. He is also the Co-Editor-in-Chief of the International Journal on Artificial Intelligence Tools and Book Series on Health Informatics. Dr. Tsai has served on the IEEE Distinguished Speaker program, US DARPA ISAT working group, and on the review panel for US NSF and NIH. He received an Engineering Foundation Research Award from the IEEE and the Engineering Foundation Society, a University Scholar Award from the University of Illinois Foundation, an IEEE Technical Achievement Award and an IEEE Meritorious Service Award from the IEEE Computer Society. He is a Fellow of the AAAS, the IEEE, and the SDPS.

Ka-Lok Ng received his PhD degree in physics from the Vanderbilt University at the US in 1990. He is a professor at the Department of Biomedical Informatics, Asia University, since August 2008. Beginning from December 2009, he serves on the Editorial board of several scientific journals. Dr. Ng has published articles in highly ranked journals, in the areas of protein interactions, robustness of protein interaction networks and microRNA.

List of Contributors

Been-Chian Chien received the B.S. degree of Computer Engineering from National Chiao Tung University in 1987, the M.S. degree and the PhD degree of Computer Science and Information Engineering in 1989 and 1992 from National Chiao Tung University, Hsinchu, Taiwan, respectively. He became an associate professor of the Department of International Trade at Nan-Tai Institute of Technology from August 1994 to July 1996 and an associate professor of the Department of Information Engineering, I-Shou University, Kaohsiung from August 1996 to July 2004. He had been invited to be a visiting scholar of the BISC (Berkeley Initiative on Soft Computing) at U.C. Berkeley, C.A. in 1999 and the School of Electrical and Computer Engineering, Georgia Institute of Technology, Atlanta, GA in 2008. In 2004, he was invited to launch and organize the Department of Computer Science and Information Engineering, National University of Tainan, Tainan, Taiwan and to be the first department head from August 2004 to July 2007 of the department. Since August 2007, he is a professor of the Department of Computer Science and Information Engineering, National University of Tainan, Tainan, Taiwan. Dr. Chien's major research interest includes computational intelligence, artificial intelligence, database systems and the design and analysis of algorithms. His current research activities involve machine learning, knowledge discovery and data mining, context-aware systems and context data management, intelligent content-based information retrieval.

Charles C. N. Wang received his M.S. and PhD degree both in bioinformatics from Asia University, Taichung, Taiwan in 2008 and 2015, respectively. He is currently a postdoctoral fellow with the department of bioinformatics, Asia University, Taichung, Taiwan. He is currently active in research related to semantic computing, natural language processing, biomedical informatics, and systems biology.

Chien-Yuan Li received his Master's Degree in Computer Science and Information Engineering in 2011 from the National University of Tainan, Taiwan. He is currently working at the Information Technology Office of National Cheng Kung University Hospital in Tainan, Taiwan. In his graduate studies, Chien-Yuan excelled in the study of cluster analysis for time series gene expression data based on autoregressive modelling approach. His expertise also includes SQL Server Administration and VB.NET.

Feng Lin is an associate professor in the School of Computer Science and Engineering, Nanyang Technological University, and the Director of Bioinformatics Research Centre. His research interest includes biomedical informatics, bioimaging, computer graphics and visualization, and high performance computing. He authored more than 200 research papers and books. He is a Senior Member of IEEE.

Hideaki Umeyama is an emeritus professor of Kitasato University, Tokyo, Japan. He received his PhD from the Pharmaceutical Department of the Tokyo University, Japan. He has also published in journals in relation to proteins including enzyme, quantum chemistry and bioinformatics such as *Proteins, Protein Eng., FEBS Lett, J. Mol. Graph. Model, Chem. Pharm. Bull. Japan*, and *J. Am. Chem. Soc.* His recent publications include comparative protein modelling and low molecular weight compound docking for protein. His current research interests include finding new drugs effective to incurable diseases such as the Hepatitis B. The awards received by Prof. Umeyama include *The Pharmaceutical Society Prize of Japan*, in 2009. 45 Publications including Hideaki Umeyama as one of the author names in the papers are searched from the URL site of http://www.pubfacts.com/author/Hideaki+Umeyama during 2002 and 2016 CE.

Hsiang-Chuan Liu received the PhD degree in Statistics from National Taiwan University, Taiwan. He is a professor at the Department of Bioinformatics, and Medical Engineering/Department of Computer Science and Information Engineering, Asia University/Graduate School of Business Administration, Fu Jen Catholic University, Taiwan, and also an honored professor at the Graduate Institute of Educational Measurement and Statistics, National Taichung UMETA-alm 1993 to 2000. Dr. Liu is a member of IEEE since 2007. He has funded research and published articles in the areas of Multivariate Analysis, Fuzzy Measure and Integral, Educational Measurement and Statistics, generalized Meta-analysis, Information Management, Machine Learning, and data mining.

Hui-Ting Lin is an assistant professor of Physical Therapy at I-Shou University (ISU), Kaohsiung, Taiwan. She received her PhD degree from Cheng Kung University, Taiwan. She has published in journals such as *Physical Therapy, Manual Therapy,* and *Clinical Biomechanics.* Her recent publications include *Effects of Pilates on Patients with Chronic Non-Specific Low Back Pain: A Systematic Review in Journal of Physical Therapy Science*, and *The Role of Negative Intra-articular Pressure in Stabilizing the Metacarpophalangeal Joint in Journal of Mechanics in Medicine and Biology.* Her current research interests include virtual reality in physical therapy application. The awards received by Prof. Lin include Excellent Advisor Award (2010), National I-Shou University.

Jinmiao Chen is a Project Leader (becoming Principal investigator on 1 April 2017) in Singapore Immunology Network, Agency for Science, Technology and Research (A*STAR), and an Adjunct Assistant Professor in the Department of Microbiology and Immunology Yong Loo Lin School of Medicine, National University of Singapore. Her interest includes single-cell RNA-sequencing data analysis, flow/mass cytometry data analysis, bioinformatics, computational biology, machine learning and data mining. She authored more than 30 journal papers.

Kung-Hao Liang is a faculty scientist in the Chang Gung Memorial Hospital, Linko, Taiwan. He has profound experience in leading multidisciplinary biomedical teams. He received his PhD from the University of Warwick, United Kingdom. He also received academic training in Academia Sinica and the Institute of Neuroscience, Queen's Medical Center, University of Nottingham in the UK. He was an adjunct assistant professor in the Life Science Department of the National Taiwan Normal University. He is the author of *Bioinformatics for biomedical science and clinical applications* (Woodhead Publishing & Elsevier, ISBN: 978-1-907568-44-2). He has published in internationally renowned journals such as the *Human Molecular Genetics, Gastroenterology and the Journal of Infectious Diseases.* His current research interests include bioinformatics, human genomics, viral hepatitis, liver diseases, rare diseases and a statistical genomics method called the generalized iterative modelling.

Mitsuo Iwadate is an associate professor of Department of Biological Sciences, Faculty of Science and Engineering, Chuo University. He completed engineering education at Oyama National College of Technology, then received his PhD from Tokyo University of Agriculture and Technology, Japan. His current research topics are protein structure modelling and in-solico screening. 23 Publications including Mitsuo Iwadate as one of the author names in the papers are searched from the URL site of http://www.pubfacts.com/author/Mitsuo+Iwadate during 2002 and 2016 CE.

Natthakan Iam-On is an assistant professor at the School of Information Technology, Mae Fah Luang University, Chiang Rai, Thailand. She received PhD in Computer Science from Aberystwyth University in 2010, funded by Royal Thai Government. Her PhD work won the Thesis Prize of 2012 by Thai National Research Council. Her present research of improving face classification for anti-terrorism and crime protection has been funded by Ministry of Science and Technology. She serves as an editor for *International Journal of Data Analysis Techniques and Strategies, International Journal of Image Mining;* as a committee and reviewer of several venues, *IEEE SMC, IEEE TKDE, Machine Learning*, for instance.

Nilubon Kurubanjerdjit received her PhD in Bioinformatics from Asia University in 2014. From 2007 to 2009, she joined the faculty at the Department of Information Technology, Kasetsart University, Thailand. Currently, she is a lecturer at the School of Information Technology, Mae Fah Luang University, Thailand. Her research interest includes PPI network, cancer-related proteins and drug discovery.

Pei-Chun Chang received his PhD degree in Chemistry from the National Taiwan University, Taiwan in 1998. Currently, Pei-Chun Chang is associate professor with the Department of Bioinformatics and Medical Engineering at Asia University where he is conducting research activities in the areas of biomedical informatics and computing chemistry. His research interests concern the discovery of cancer biomarkers by genomics analysis, the screening of anticancer compounds from Chinese herb components, the development of disease model with systems biology, and medical data mining.

Pei-Lin Chen received her Master's Degree in Computer Science and Information Engineering in 2014 from the National University of Tainan,

Taiwan. She is currently working at the Kao-Ching Ranch Farm in Pingtung, Taiwan. In her graduate studies, Pei-Lin excelled in the study of cluster analysis for time series gene expression data based on support vector machine approach. Her expertise also includes developing customized software using Matlab.

Phillip C.-Y. Sheu received his PhD degree in electrical engineering and computer science from the University of California, Berkeley in 1986. He is a professor of EECS, CS, and BME at the University of California, Irvine, and a guest professor with the Department of Biotechnology and Bioinformatics, Asia University, Taiwan. Dr. Sheu is a Fellow of IEEE. He is currently active in research related to semantic computing, robotic computing, biomedical computing and multimedia computing.

Rong-Ming Chen received his PhD degree in Electrical Engineering from the National Tsing Hua University, Hsinchu, Taiwan. He is a professor of Computer Science and Information Engineering at the National University of Tainan, Taiwan. His current research interests include bioinformatics, signal processing, parameter estimation and data science.

Rouh-Mei Hu received her PhD degree from the Université Paris-Sud (Paris-XI), Orsay, France. She is a professor of Bioinformatics and Medical Engineering at Asia University, Taiwan. Her current research interests include transcriptional regulation in microorganism and human microbiomes.

Tossapon Boongoen is an associate professor at Mae Fah Luang University, Thailand. He obtained PhD in Artificial Intelligence from Cranfield University in 2003, and worked as a Post-Doctoral Research Associate (PDRA) at Aberystwyth University, during 2007–2010. His PDRA work focused on anti-terrorism using data analytical and decision support synthesizes. He has been the leader of research projects in exploiting biometrics technology for anti-terrorism in southern-provinces of Thailand, funded by Ministry of Defense. He has been an editor of several international journals such as *AMB Express* and *International Journal of Collaborative Intelligence*. Also, he serves as a committee and reviewer of several venues, IEEE SMC, IEEE TKDE, Knowledge Based Systems, *International Journal of Intelligent Systems Technologies and Applications*, for instance.

Wen-Pin Hu is currently an associate professor in the Department of Bioinformatics and Medical Engineering at Asia University, Taiwan. He received his B.S. degree in Mechanical Engineering from National Sun Yet-sen University in Kaohsiung City, Taiwan. He completed M.S. and PhD degrees in the Institute of Biomedical Engineering at National Cheng Kung University in Tainan City of Taiwan. He had been a visiting scholar at University of Washington, Seattle, USA, from August 2005 to March 2006. His current research interests include biosensing technologies, aptasensor, simulation of protein-nucleic acid interaction and biomechanics.

Wen-Yih Chen is currently a distinguished professor in the Department of Chemical and Materials Engineering and Institute of Biomedical Engineering, National Central University (NCU). He was the Chairman of the Department and was the Associated Dean of the Engineering School of NCU. He also was a visiting professor of MIT, Monash University, and University of Washington, Seattle. His research emphases have been on understanding the thermodynamics and kinetics of biomolecular interactions. With the principle understanding of the molecular interactions, he has successfully elucidated some biochemical separations mechanism, protein folding disease phenomena and development of antifouling materials in molecular level. Currently, he has devoted his research resources in biosensor development, especially in gene sequencing and biomarkers detection by Field Effect Transistor and Surface Plasmon Resonance. He has published more than 150 peer review papers and owned and applied more than 40 patents, mostly in biosensor area. He is currently on the editorial board of Biotechnology Journal (SCI), Bioprocess and Bioengineering (SCI) and Atomic and Molecular Physics (SCI).

Yoshiki Murakami is an associate professor of Department of Hepatology, Graduate School of Medicine, Osaka City University. He received medical education at Kanazawa University, then received his PhD from the Kyoto Prefectural University of Medicine, Japan. His current research topics are nucleic acid drug discovery and biomarker by using circulating miRNA. He has written 52 articles and 13 books to date.

Y-h. Taguchi is professor of Department of Physics at Chuo University, Tokyo, Japan. He received his Dr. Sci. from the Tokyo Institute of Technology, Tokyo, Japan. He has also published in journals such as the PLoS One and BMC Bioinformatics. His recent publications include *Principal component analysis based unsupervised feature extraction applied to publicly available gene expression profiles provides new insights into the* *mechanisms of action of histone deacetylase inhibitors* in Neuroepigenetics. His current research interests include principal component analysis as well as tensor decomposition based feature extraction and its application to Bioinformatics. http://orcid.org/0000-0003-0867-8986.

Contents

Chapter 1

Unsupervised clustering of time series gene expression data based on spectrum processing and autoregressive modeling[a]

Chien-Yuan Li, Rong-Ming Chen* and Been-Chian Chien
Department of Computer Science and Information Engineering, National University of Tainan, Taiwan

Rouh-Mei Hu and Jeffrey J. P. Tsai
Department of Bioinformatics and Medical Engineering, Asia University, Taiwan

1.1 Introduction

With the growing value of advances in biotechnology, bioinformatics has become a promising discipline. Bioinformatics research integrates the fields of biology, molecular biology, medicine, pharmacology, information science, mathematics, physics, and chemistry to study topics such as comparing and analyzing similarities between genes and proteins, gene mapping, protein structural analysis, and evolutionary relationships of genes. For example, suppose genetic analysis is necessary to determine whether a disease is genetic or congenital, but the amount of genetic data is excessive. File comparison software such as NBCI BLAST (http://www.ncbi.nlm.nih.gov) or self-written algorithms

---

[a] An earlier version of this study in Chinese was presented at The Conference on Technologies and Applications of Artificial Intelligence (Domestic Track), Taiwan, Nov. 11-13, 2011.

* Corresponding author.

can be used to analyze this data. In addition, bioinformatics can provide computational methods for the effective categorization or analysis of big data generated by large-scale experiments.

DNA microarray (also called gene chip) technology was first developed by the Stanford University Biochemistry Department [11]. This technology rapidly advanced the study of gene expression to enable simultaneous observation of up to tens of thousands of gene expressions. Microarrays are primarily used in 10 research areas, including biology, pharmacology, and molecular biology [5]. Time series gene expression data is obtained by measuring DNA microarray probes at various times.

Two types of gene expression data exist: steady state and time series gene expression. Steady state gene expression data is obtained by recording the values of one gene expression from varying entities, tissues, or experiments. Time series gene expression data is obtained by recording the values of various genes at varying time points. The cluster analysis of time series gene expressions explored in this study used time series gene expression data. The prohibitive cost of DNA microarray technology results in temporal data with few time points. The purpose of cluster analysis of time series gene expressions is to determine an effective clustering algorithm that groups genes with similar temporal patterns into a cluster. Similarities in temporal patterns among genes may represent similarities in biological significance on a biomolecular level or participation in the same biological process, from which researchers can infer that these genes belong to the same genetic regulatory network.

Numerous previous studies have proposed techniques for improving clustering results based on characteristics of the data being analyzed. However, when these techniques were tested on real-life experimental biological data, such as time series gene expression data, performance evaluation metrics indicated much room for debate and improvement [6, 11, 30]. Therefore, this study proposed a new mixed data processing model that integrates spectrum processing and autoregressive modeling. Spectrum processing can resolve time displacement in a time series. Autoregressive modeling exhibits excellent expression of the dynamic behavior of real-life biological time series data. The researchers hoped that the results of clustering time series gene expression data could be improved by first using spectrum processing followed by autoregressive modeling to express dynamic data. Finally, the results of the technique proposed in this study were evaluated for validity and analyzed for statistical significance as well as correlations with biological significance based on gene ontology. The results were also compared with three frequently used traditional clustering algorithms: k-means [14, 24], hierarchical [12], and self-organizing map (SOM) [23].

The remaining sections of this chapter are organized as follows: Section 1.2 contains the literature review, which summarizes other analyses of time series gene expression and the characteristics and limitations of some popular clustering algorithms used to examine the results of the proposed algorithm. Section 1.3 details the methods and introduces the proposed data processing algorithm, which involves preprocessing of the data, spectrum processing [3], singular value decomposition [4], and autoregressive modeling [9]. Section 1.4 presents the results and includes a discussion of the experimental data used in this study as well as the evaluation metrics and results. Next, the proposed data processing method is validated and compared with three frequently used traditional algorithms; the statistical significance of these results is discussed. Lastly, the clustering results are explained in terms of gene annotation, based on the semantics of gene ontology. The conclusion of this study and suggestions for future research are presented in Section 1.5.

1.2 Literature Review

This section introduces studies and applications of time series gene expression cluster analysis and explores three frequently used traditional clustering algorithms that are compared to the data processing method proposed in this study: k-means, hierarchical, and SOM. An extensive collection of domestic and international studies and their theoretical bases was compiled to understand topics related to this study. Besides, since the Chinese restaurant clustering (CRC) algorithm is able to infer number of clusters, this study used the CRC algorithm to determine the reference number of clusters for the clustering algorithms, the CRC will also be introduced briefly.

1.2.1 *Cluster Analysis of Time Series Gene Expression*

Multiple previous studies related to clustering problems in time series gene expression data merit description. For example, Eisen *et al.* [12] in 1998 applied the hierarchical average-linkage clustering algorithm to budding yeast genes. Tavazoie *et al.* [24] separately applied SOM and k-means clustering algorithms to gene expression data and determined motifs related to upstream DNA sequences and coexpressed genes in 1999. Yeung and Ruzzo [29] in 2001 made an empirical study on principal component analysis for clustering gene expression data. In 2004, Zhou *et al.* [31] proposed a clustering method based on minimizing mutual information among clusters, integrating a heuristic search algorithm to resolve the optimization problem.

Numerous methods for converting raw data through spectrum transform exist. For example, Spellman *et al.* [22] used Fourier transform on asynchronous data. If a time series does not consist of fixed time intervals, data without fixed time intervals may result. Lack of fixed time intervals may be caused by incomplete data or changed sampling frequency. Therefore, Zhao *et al.* [30] used a Lomb–Scargle periodogram to analyze the gene expression data of these types of time series, calculate coexpressed values, and compare various clusters using varying distance measurement methods [16, 21]. Darvish *et al.* [9] first used k-means to achieve gene clustering, and then used autoregressive modeling to modularize correlations between all genes so that the next time point could be predicted. This is an excellent solution for expression data without sufficient time points. In most situations, autoregressive modeling requires large amounts of training data to determine the coefficients, but cost concerns limit the number of time points. Therefore, Darvish *et al.* [10] first used nonlinear component analysis on gene expression data to find the primary components of gene clusters before using autoregressive modeling analysis. This produced results close to true values.

Targeting circumstances in which genes can simultaneously be assigned to more than one appropriate cluster, Bandyopadhyay *et al.* [2] proposed using a two-stage clustering algorithm for significant multiclass membership (SiMM-TS). In the first stage, a variable string length genetic algorithm [17] was used to determine the number of clusters and identify significant multiclass membership (SiMM) genes. In the second stage, a multiobjective genetic algorithm [1] was used to cluster the genes after SiMM genes were filtered. Finally, the SiMM genes were assigned to one of the clusters defined in the second stage based on the nearest neighbor criterion.

1.2.2 *Clustering Algorithm*

1.2.2.1 *K-Means*

The k-means clustering algorithm requires predefining the number of clusters. Assume n clusters at the beginning. The algorithm then randomly chooses n starting points from among the data and computes the distance between each remaining data point and the starting points to determine the cluster to which the data point should be assigned. This study used the Pearson's correlation to determine the distance between points [15, 25]. After performing calculations to determine the number of groups, the centroids of n groups were recalculated. If new cluster centroids were found, the new centroids were used to recalculate the distance between all data points and the new centroids. If not, the clustering results were considered stable and consistent.

1.2.2.2 *Hierarchical*

Hierarchical clustering algorithms begin at the bottom layer of a dendrogram, and are integrated layer by layer. At the beginning, each point is considered a cluster. Assuming n points, these n points become n clusters; in other words, each cluster contains one point. The two clusters separated by the shortest distance are then identified, named C_i and C_j, and combined into one new cluster. The two closest clusters are again identified and combined, and this process repeats until the number of clusters matches the predetermined number. When determining the two closest clusters, single-linkage clustering often results in a minimum spanning tree, which causes an undesirable long chain. Complete linkage clustering does not result in the chaining phenomenon, but only produces valid results when the clusters are compact. Therefore, average-linkage clustering is typically used to calculate the closest distance, which is an average of the single-linkage and the complete-linkage. As with k-means, the Pearson's correlation is used to determine the distance between two points.

1.2.2.3 *SOM*

SOM is a type of artificial neural network, and an unsupervised learning network. The guiding principle is the emulation of neurons in the cerebrum, where neurons with similar functions converge. An SOM primarily consists of three stages: input, output, and weighting. The features of the data are used as training data. Through repeated training, all data are mapped to nodes. This mapping is dependent on a feature vector and the degree of similarity of the data; data closest to a node is highest in similarity. The advantage of an SOM is that it can simplify high-dimensional data; the disadvantage is that an SOM is relatively slow because the network must be trained.

1.2.2.4 *CRC*

Chinese restaurant clustering, as the name implies, is a clustering algorithm analogous to seating customers at a Chinese restaurant. The algorithm is based on the Chinese restaurant process (CRP), and was therefore named CRC. The algorithm supposes that a Chinese restaurant that can accommodate an infinite number of tables, and each table can seat an unlimited number of customers. A customer who enters the restaurant can choose to sit at either an empty table or at a table with other customers. Applying this clustering algorithm to gene clustering, the customers are genes, and tables are clusters. The algorithm is able to infer number of clusters. The functions for calculating cluster probabilities and updating clusters are not detailed in this study; please refer to the original study [19].

1.3 Methods

This section details the methods and introduces the proposed data processing algorithm, which includes the preprocessing of the time series gene expression data, spectrum processing, singular value decomposition, and autoregressive modeling.

1.3.1 *Data Processing Algorithm*

The underlying concept of the proposed algorithm is the integration of spectral similarity and autoregressive modeling data processing methods. Time series gene expression data was first converted to a spectrum. The resulting gene versus spectrum array resolved the problem of displacement. Singular value decomposition was then used to reduce the size of the gene versus spectrum array and to identify the optimal orders of autoregressive modeling. Autoregressive modeling was then used to estimate all model coefficients of each gene expression series. These model coefficients were used to calculate the Pearson's correlation. The pairwise correlation coefficient between two genes was then used as the distance measure for cluster analysis. Figure 1.1 shows the flowchart for the proposed data processing algorithm for clustering of time series gene-expression data based on spectrum processing and autoregressive modeling.

1.3.2 *Data Preprocessing*

Based on background noise variables, genes with gene expression values that do not vary significantly are usually excluded from analysis [2]. In addition, time series gene expressions with missing values were not considered in this study. If the data contains time series gene expressions of genes with the same name, these gene expressions be merged, and the typical way of handling this problem is to use the average of the gene expression values. For part of the time series gene expression data, only genes with more obvious expression values were extracted; this is explained in more detail in Section 1.4.2. Finally, each time series gene expression record was Z-normalized so that the average gene expression value of each row was 0 and the variance was 1.

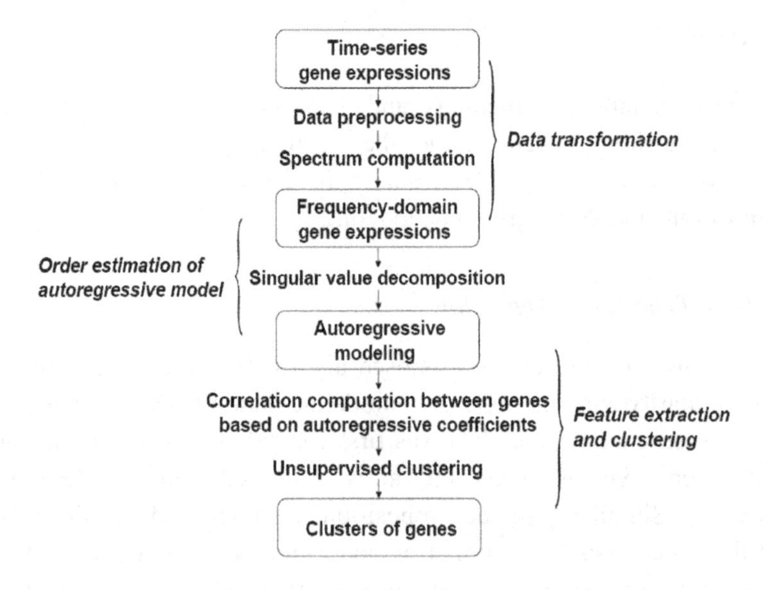

Fig. 1.1. Flowchart for the proposed data processing algorithm for clustering of time series gene-expression data based on spectrum processing and autoregressive modeling.

1.3.3 *Spectrum Transform*

First, suppose that each gene g has N time points. The time series gene expression can be expressed as

$$x_g = \left[x_g(0)\, x_g(1) \cdots x_g(N-1) \right]^T \tag{1.1}$$

where $x_g(n)$, $n = 0, 1, ..., N-1$ represents the gene expression values of gene g at time n. Discrete Fourier transform is used to convert the time series into a spectrum, as follows:

$$X_g(k) = \frac{1}{N} \sum_{n=0}^{N-1} x_g(n) e^{-j*2\pi*n*k/N} \tag{1.2}$$

where $j = \sqrt{-1}$ and $k = 0, 1, ..., N-1$. The information represented by the Fourier coefficient $\{X_g(k)\}$ is a function of k. Because k is positively correlated to the frequency, these coefficients $\{X_g(k)\}$ are frequently referred to as a spectrum.

1.3.4 *Singular Value Decomposition*

Suppose that spectrum processing results in an $M \times N$ sized genes (rows) versus spectrum (columns) matrix A. Singular value decomposition is used to determine the appropriate order for autoregressive modeling. First, matrix A is decomposed to $A = U\Sigma V^T$, where Σ is the diagonal matrix of an $M \times N$ array, and the diagonal value is a nonnegative real number. The value that immediately precedes the diagonal value approaching zero is the order number for autoregressive modeling.

1.3.5 *Autoregressive Modeling*

The main advantage of autoregressive modeling is that it can represent the dynamic behavior of a data series. An autoregressive model of order p is defined as follows [9]:

$$x(t) + a_1 x(t-1) + \ldots + a_p x(t-p) = e(t) \tag{1.3}$$

where $x(t), x(t-1), \ldots, x(t-N+1)$ is a time series of length N, $a_1, \ldots, a_p$ is the autoregressive coefficient, and $e(t)$ is white noise. In addition, $\theta = [a_1, a_2, \ldots, a_p]$ is defined as the parameter vector, and $\varepsilon(t, \theta)$ is the estimated error:

$$\varepsilon(t, \theta) = x(t) - \hat{x}(t \mid \theta) \tag{1.4}$$

where $\hat{x}(t \mid \theta)$ is the estimated output calculated from the vector θ. The least squares error method was then used to estimate these parameters. First, assume

$$\hat{x}(t \mid \theta) = \varphi^T(t)\theta \tag{1.5}$$

where φ is the regression vector, defined as

$$\phi(t) = [-x(t-1) \; -x(t-2) \; \ldots \; -x(t-p)]^T \tag{1.6}$$

Substituting (1.5) into (1.4) results in

$$\varepsilon(t, \theta) = x(t) - \varphi^T(t)\theta \tag{1.7}$$

Using the least squares error method produces

$$\hat{\theta} = \left[\frac{1}{p} \sum_{t=1}^{p} \varphi(t)\varphi^T(t) \right]^{-1} \left[\frac{1}{p} \sum_{t=1}^{p} \varphi(t)x(t) \right] \tag{1.8}$$

After estimating the coefficient vector of the autoregressive model, this vector can be used to calculate the correlation coefficient between two genes in a gene expression series.

1.4 Experimental Results and Analysis

This section reveals the experimental results of the proposed data processing algorithm. The algorithm was applied to and compared with three frequently used traditional clustering algorithms. The experimental data comprised five sets of real-life biological data published in past studies and one set of synthetic data.

1.4.1 *Evaluation Methods*

To validate the proposed data processing algorithm, the silhouette index, a commonly used evaluation metric to calculate the tightness of clusters after clustering was used [20]. Subsequently, the Wilcoxon rank-sum test was used to determine the statistical significance of the results [13]. Lastly, the Kyoto Encyclopedia of Genes and Genomes (KEGG), which is a database of genetic regulatory networks, was used to describe the correlation between clustering results and their biological significance. The evaluation metrics used in this study are briefly introduced below.

1.4.1.1 *Silhouette Index*

The silhouette index evaluates cluster validity using the distance between points in various clusters and the distance between all points in all clusters. This study used the Pearson's correlation to determine the distance between points. The silhouette index is defined as the average silhouette width of all points (genes) in all clusters. In Fig. 1.2, which shows a schematic diagram of silhouette width, *A*, *B*, and *C* are three clusters. Assuming point i is a point in cluster A, the average distance from i to all other points in cluster A is a, and the shortest average distance from i to other clusters is b. The equation to determine the silhouette width for point i is defined as

$$s(i) = \frac{b - a}{\max(a, b)} \tag{1.9}$$

The value of $s(i)$ can range from –1 to 1; values close to 1 imply that i has been assigned to an appropriate cluster.

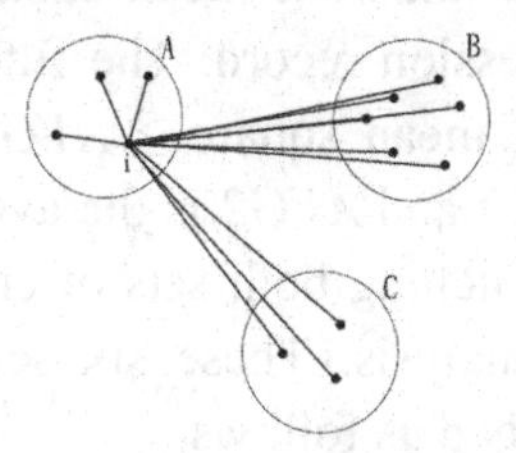

Fig. 1.2. Schematic diagram for calculating silhouette width.

1.4.1.2 *Wilcoxon Rank-Sum Test*

The Wilcoxon rank-sum test is a non-parametric statistical hypothesis test which can be used to infer differences between the two populations by comparing the difference in median values between two random samples. The null hypothesis is that the two samples are derived from populations with similar characteristics. The data measurement is based on an ordinal scale. The advantages of the Wilcoxon rank-sum test are that the test is simple to calculate and is not easily affected by extreme values.

1.4.2 *Experimental Data*

This study used the same experimental data as [2], [9], and [21]. A total of six sets of time series gene expression data were tested, including one set of synthetic data (named AD400_10_10) and five sets of real-life experimental biological data. The five sets of biological data included one set of data each from human fibroblast serum, rat central nervous system (CNS), yeast sporulation, and two sets of data from *Saccharomyces cerevisiae*; all data sets are available for download online. Emulating previous studies in the treatment of data sets with high numbers of genes (i.e., yeast sporulation, *S. cerevisiae* [7], and *S. cerevisiae* [22]), only genes with time series gene expression values that varied significantly were used in this study; thus, the regulatory relationships between two genes were readily apparent [27]. In this study, genes with time series gene expression values that did not vary significantly or genes that did not significantly differ in gene expression were filtered. The variance and root mean square were calculated for each time series gene expression record. The filter threshold was set to variance $\geq$ *AVG1* and root mean square $\geq$ *AVG2*, where *AVG1* is the average variance of all genes and *AVG2* is the average root mean square of all genes. Only genes matching both sets of criteria were included in the data sets for further analysis. These six sets of time series gene expression data were described as follows.

1.4.2.1 *Dataset #1: AD400_10_10 (Bandyopadhyay et al. [2])*

This time series gene expression data set contains synthetic data, comprising 400 genes measured at 10 time points. The data set contains 10 clusters, and each cluster contains 40 genes; in other words, the data contains 10 vastly dissimilar time series gene expression styles [28].

1.4.2.2 *Dataset #2: Human Fibroblasts Serum (Eisen et al. [12])*

This set of real-life experimental biological experimental data contains 517 genes measured at 13 time points.

1.4.2.3 *Dataset #3: Yeast sporulation (Chu et al. [8])*

This real-life experimental biological data set contains the raw data of 6118 genes measured at 7 time points. Prior to cluster analysis, the previously described method was used to filter the genes of which the gene expression values did not vary significantly. This filtering resulted in a subset of 844 time series gene expression values that were used in this experiment.

1.4.2.4 *Dataset #4: Rat CNS (Wen et al. [26])*

This real-life biological data set contains 112 genes measured at 9 time points.

1.4.2.5 *Dataset #5: S. cerevisiae (Cho et al. [7])*

This real-life experimental biological data set contains 6239 genes measured at 17 time points; this time series gene expression data set can be obtained from the following Web site: http://yfgdb.princeton.edu/ cgi-bin/display.cgi?id=9702192&db=pmid. In this study, genes whose expression values did not vary significantly were filtered, resulting in a subset of 1193 time series gene expression values that were used for cluster analyses.

1.4.2.6 *Dataset #6: S. cerevisiae (Spellman et al. [22])*

This real-life experimental biological data set contains 8832 genes measured at 24 time points. For this study, genes whose expression values did not vary significantly were filtered, resulting in a subset of 1380 time series gene expression values that were used in cluster analyses.

1.4.3 *Experimental Results*

This section introduces the experimental parameters settings for the clustering algorithms used in this study and the estimation of the orders for autoregressive modeling. The use of the silhouette index and

Wilcoxon rank-sum test described in the previous section to evaluate experimental results is explained. Because several algorithm initialization values were random, and these values affected the final clustering results, each algorithm was executed 10 times on all data sets. The median value was used to calculate the silhouette index, and the Wilcoxon rank-sum test was performed on this silhouette index to calculate the p value.

1.4.3.1 *Experimental Parameter Settings*

For k-means and hierarchical algorithms, the Pearson's correlation was used to calculate distance. The minimal distance for the hierarchical algorithm was defined using average-linkage. For the SOM algorithm, preset Matlab parameter values were used. For CRC algorithm parameter settings, system presets were preserved for *num_chains* and *num_cycles*. The parameter *inversion_flag* was set to 0, indicating that gene expression could not be both synchronous and opposite trending, and *max_shift* was set to 0, indicating that time shifts were not considered. Based on experimental testing, these settings resulted in optimal values. The parameter *prob_cutoff* was also set to 0, indicating that no threshold limit existed for the probability of a gene belonging to a cluster; this avoided the removal of genes because of the inability to be assigned to a cluster.

Because most clustering algorithms require the predefinition of the number of clusters, and the CRC algorithm is able to infer number of clusters, this study used the CRC algorithm to first cluster the experimental data and obtain the cluster number with the highest silhouette index, and used this as the reference cluster number for each set of data. Applying this method to the human fibroblast serum, yeast sporulation, and rat CNS data sets yielded cluster numbers of 10, 10, and 4, respectively. The cluster number for the synthetic data set AD400_10_10 was already known to be 10. Lastly, according to the KEGG Web site[b], the two sets of data for *S. cerevisiae* pertained to the

[b] http://www.genome.ad.jp/kegg/pathway/sce/sce04111.html

cell cycle phases G1, S, G2, and M; therefore, the number of clusters for these two data sets was four.

1.4.3.2 *Analysis and Comparison of Experimental Results*

Previous studies have explored using spectrum processing [30] or autoregressive modeling [27] in time series gene expression clustering; however, unlike the present study, none have combined both data processing methods to clustering algorithms. This study compared the results of four data processing methods applied to three frequently used traditional clustering algorithms. The four methods were raw data, spectrum processing, autoregressive modeling, and a combination of spectrum processing and autoregressive modeling. The three clustering algorithms were k-means, hierarchical, and SOM. The experimental data comprised the six sets of time series gene expression data that were described previously.

Table 1.1 lists the median silhouette index values obtained after 11 executions of each clustering algorithm. Comparing the silhouette index values in Table 1.1 shows that the clustering results obtained by the proposed data processing method were more valid than were those obtained from the other three data processing methods for the AD400_10_10, human fibroblasts serum, and yeast sporulation data sets. The proposed data processing method was less valid than were pure spectrum processing and pure autoregressive modeling in only two instances. In the hierarchical clustering of the rat CNS data set and the *S. cerevisiae* data set of Cho *et al.*, pure autoregressive modeling was more valid than was the proposed data processing method; however, the proposed method was more valid than was using the raw data or pure spectrum processing. In k-means clustering of the *S. cerevisiae* data set of Spellman *et al.*, spectrum processing alone was more valid than was the proposed data processing method; however, the proposed method was more valid than was using the raw data or pure autoregressive modeling. The reasons for these results may be related to the distribution of data or the initialization settings of the clustering algorithms. However, as shown in Table 1.1, executing the three clustering algorithms on the six data sets using the proposed data processing method as well as three other methods resulted in 54 instances. The proposed method was more valid than were the other three methods in 51 of those instances (approximately 94.4% of the time). Thus, in the majority of instances, the proposed data processing method was more valid.

Although the proposed data processing method in clustering algorithms resulted in higher median silhouette index values in the majority of experimental tests compared with the other methods, the significance of these results required statistical verification. Therefore, the 11 silhouette index values obtained for each algorithm were used to calculate p values using the Wilcoxon rank-sum test. A p value less than .05 indicated that the differences in the two populations were statistically significant.

The values displayed in parentheses in Table 1.1 are the p values obtained by applying the Wilcoxon rank-sum test to the 11 pairs of silhouette index values of the four data processing methods. Table 1.1 shows that in most instances, the difference between raw data and gene expression data that was processed using both spectrum processing and autoregressive modeling was statistically significant. For these populations, the p value was higher than .05 in only two instances: the *S. cerevisiae* data of Cho *et al.* in the SOM algorithm and the *S. cerevisiae* data of Spellman *et al.* in the k-means algorithm. This indicated that although the median silhouette index value obtained from raw data was higher than was that obtained from gene expression data that was processed using both spectrum processing and autoregressive modeling, no strong statistical significance existed to support this comparison. Comparison of raw data and gene expression data that was processed using only autoregressive modeling resulted in only one instance where the p value was higher than 0.05: the *S. cerevisiae* data of Cho *et al.* in the k-means algorithm. In all other instances, the comparison results were statistically significant.

Table 1.1. Median silhouette index values obtained after 11 executions of the clustering algorithms. Values marked with a (*) are higher than the value obtained by the proposed data processing method. The bottom values in parentheses indicate the p values calculated by comparing the specified method and the proposed method (combining both spectrum processing and autoregressive modeling).

Algorithm / Data processing	Dataset / Number of clusters					
	#1/10	#2/10	#3/11	#4/4	#5/4	#6/4
K-means / Original data	0.2697 (7.8e-5)	0.3095 (7.8e-5)	0.3898 (7.8e-5)	0.4415 (7.8e-5)	0.2789 (7.8e-5)	0.2267 (0.205)
K-means / Data with spectrum processing [30]	0.3102 (7.8e-5)	0.2622 (7.8e-5)	0.5061 (7.8e-5)	0.505 (0.0068)	0.3366 (0.358)	0.2734* (0.009)
K-means / Data with AR modeling [9]	0.4475 (7.8e-5)	0.5179 (7.8e-5)	0.5179 (7.8e-5)	0.4659 (0.0017)	0.2804 (7.8e-5)	0.1875 (7.8e-5)
K-means / The proposed method	**0.4962**	**0.38**	**0.5955**	**0.5647**	**0.3367**	**0.233**
Hierarchical / Original data	0.3441 (7.8e-5)	0.2662 (7.8e-5)	0.3566 (7.8e-5)	0.3894 (7.8e-5)	0.2151 (7.8e-5)	0.1233 (7.8e-5)
Hierarchical / Data with spectrum processing [30]	0.2748 (7.8e-5)	0.2126 (7.8e-5)	0.4561 (7.8e-5)	0.4174 (7.8e-5)	0.211 (7.8e-5)	0.1261 (7.8e-5)
Hierarchical / Data with AR modeling [9]	0.3481 (7.8e-5)	0.2693 (7.8e-5)	0.4448 (7.8e-5)	0.4797* (7.8e-5)	0.2686* (7.8e-5)	0.0977 (7.8e-5)
Hierarchical / The proposed method	**0.4281**	**0.3312**	**0.4955**	**0.4377**	**0.2593**	**0.1529**
SOM / Original data	0.1775 (7.8e-5)	0.2018 (7.8e-5)	0.131 (7.8e-5)	0.3297 (7.8e-5)	0.1775 (1.4e-4)	0.1579 (7.8e-5)
SOM / Data with spectrum processing [30]	0.0452 (7.8e-5)	0.0053 (7.8e-5)	0.1493 (7.8e-5)	0.1405 (7.8e-5)	0.1158 (5.7e-4)	0.0874 (7.8e-5)
SOM / Data with AR modeling [9]	0.1946 (7.8e-5)	0.1137 (7.8e-5)	0.0946 (7.8e-5)	0.2943 (7.8e-5)	0.14 (7.8e-5)	0.1025 (7.8e-5)
SOM / The proposed method	**0.3852**	**0.2567**	**0.4712**	**0.4106**	**0.1852**	**0.1684**

Table 1.2. Number of times that the SIC1 and CLB2 genes were assigned to the same cluster using the S. *cerevisiae* experimental data of Spellman *et al.*. The number of clusters was set to four. Three algorithms using four data processing methods were executed 11 times each.

		Data processing			
		Original data	Data with spectrum processing [30]	Data with AR modeling [9]	The proposed method
Algorithm	K-means	0	11	9	**11**
	Hierarchical	0	11	11	**11**
	SOM	0	11	0	**11**

1.4.4 *Analysis of Correlation with Biological Significance*

In this section, we analyze the clustering of the SIC1 and CLB2 genes in the *S. cerevisiae* data set of Spellman *et al.* based on the *S. cerevisiae* cell cycle genetic regulatory pathway (KEGG) to explore the clustering results of these two genes when the proposed data processing method and the other three methods were applied. SIC1 and CLB2 genes are expressed in the G2 stage of the cell cycle [18]. The four data processing methods were applied to the three clustering algorithms, with the number of clusters set to four. Each combination of method and algorithm was executed 11 times. The number of times that SIC1 and CLB2 were assigned to the same cluster is recorded in Table 1.2. This information indicated that SIC1 and CLB2 were assigned to the same cluster for all 11 executions of the clustering algorithms, signifying that the proposed data processing method could effectively express correlations based on biological significance.

1.5 Conclusion and Future Research Directions

This chapter explored clustering problems in time series gene expression, primarily considering time displacement in regulatory relationships between genes. Spectrum processing was used to correct the tendency for these genes to be identical. Next, autoregressive modeling, which exhibits excellent expression of the dynamic behavior of data, was incorporated, and the autoregressive modeling coefficients were used as feature vectors. Clustering results significantly improved. Numerous previous studies have proposed techniques to improve clustering results based on characteristics of the data being analyzed. However, when these techniques were tested on real-life biological data, such as time series gene expression data, performance evaluation metrics exhibited much room for debate and improvement.

The clustering of time series gene expression data is a highly researched topic. Scholars have used varying data preprocessing methods, improvements to clustering algorithms, and various evaluation techniques to explain their research; therefore, comparing the various results is difficult. Although the proposed data processing method produced results that appeared ideal, future validation of this method should compare the method to additional algorithms to achieve a more objective evaluation. For example, the adjusted Rand index [29] could be used to validate clustering results. However, using this evaluation method requires prior knowledge of the correct results in time series gene expression data. In addition, integrating knowledge of the considered domain into the algorithm is another research direction to ensure highly significant correlations between clustering results and biological significance.

Acknowledgments

This work was supported in part from the Ministry of Science and Technology, Taiwan [Grant Numbers: NSC 100-2221-E-024-020 and MOST 103-2221-E-024-014].

References

1. Bandyopadhyay, S., Maulik, U. and Mukhopadhyay, A. (2007). Multiobjective genetic clustering for pixel classification in remote sensing imagery, IEEE Transactions on Geoscience and Remote Sensing, 45, pp. 1506–1511.
2. Bandyopadhyay, S., Mukhopadhyay, A. and Maulik, U. (2007). An improved algorithm for clustering gene expression data, Bioinformatics, 23, pp. 2859–2865.

3. Bracewell, R. (1999). *The Fourier Transform and Its Applications*, 3rd Ed. (McGraw Hill, USA).
4. Bretscher, O. (1997). *Linear Algebra with Applications*, (Prentice Hall, USA).
5. Chen, J. J. W. (2000). Introduction and application of DNA microarrays: A formidable weapon for genetic analysis in the 21st century, NTU BioMed Bulletin, 2, pp. 18–25 (in Chinese).
6. Chiu, T. Y., Hsu, T. C., Yen C. C. and Wang, J. S. (2015). Interpolation based consensus clustering for gene expression time series, BMC Bioinformatics, 16, pp. 117–133.
7. Cho, R. J., Campbell, M. J., Winzeler, E. A., Steinmetz, L., Conway, A., Wodicka, L., Wolfsberg, T. G., Gabrielian, A. E., Landsman, D., Lockhart, D. J. and Davis, R. W. (1998). A genome-wide transcriptional analysis of the mitotic cell cycle, Molecular Cell, 2, pp. 65–73.
8. Chu, S., DeRisi, J., Eisen, M., Mulholland, J., Botstein, D., Brown and P. O., Herskowitz, I. (1998). The transcriptional program of sporulation in budding yeast, Science, 282, pp. 699–705.
9. Darvish, A., Hakimzadeh, R. and Najarian, K. (2004). Discovering dynamic regulatory pathway by applying an auto regressive model to time series DNA microarray data, *Proc. 26th Annual International Conference of the IEEE Engineering in Medicine and Biology Society*, IEMBS, pp. 2941–2944.
10. Darvish, A., Najarian, K., Jeong, D. H. and Ribarsky, W. (2005). System identification and nonlinear factor analysis for discovery and visualization of dynamic gene regulatory pathways, *Proc. IEEE Symposium on Computational Intelligence in Bioinformatics and Computational Biology*, CIBCB, pp. 1–6.
11. DeRisi, J., Penland, L., Brown, P. O., Bittner, M. L., Meltzer, P. S., Ray, M., Chen, Y., Su, Y. A. and Trent, J. M. (1996). Use of a cDNA microarray to analyze gene expression patterns in human cancer, Nature Genetics, 14, pp. 457–460.
12. Eisen, M. B., Spellman, P. T., Brown, P. O. and Botstein, D. (1998). Cluster analysis and display of genome-wide expression patterns, *Proc. National Academy of Sciences of the United States of America*, 95, pp. 14863–14868.
13. Hollander M. and Wolfe, D. A. (1999) *Nonparametric Statistical Methods*, 2nd Ed. (Wiley, USA).
14. Jain, A. K. (2010). Data clustering: 50 years beyond K-means, Pattern Recognition Letters, 31 pp. 651–666.
15. Jaskowiak, P. A., Campello, R. J. and Costa, I. G. (2014). On the selection of appropriate distances for gene expression data clustering, BMC Bioinformatics, 15, suppl. 2, pp. S2–S18.
16. Lomb, N. R. (1976). Least-squares frequency analysis of unequally spaced data, Astrophysics and Space Science, 39, pp. 447–462.
17. Maulik U. and Bandyopadhyay, S. (2002). Performance evaluation of some clustering algorithms and validity indices, IEEE Transaction on Pattern Analysis and Machine Intelligence, 24, pp. 1650–1654.

18. Noguchi, E. and Gadaleta, M. C. (2014). *Cell Cycle Control: Mechanisms and Protocols*, (Humana Press, USA).
19. Qin, Z. S. (2006). Clustering microarray gene expression data using weighted Chinese restaurant process, Bioinformatics, 22, pp. 1988–1997.
20. Rousseeuw, P. (1987). Silhouettes: a graphical aid to the interpretation and validation of cluster analysis, Journal of Computational and Applied Mathematics, 20, pp. 53–65.
21. Scargle, J. D. (1982). Studies in astronomical time series analysis. II — Statistical aspects of spectral analysis of unevenly spaced data, Astrophysical Journal, Part 1, 263, pp. 835–853.
22. Spellman, P. T., Sherlock, G. and Zhang, M. Q. (1998). Comprehensive identification of cell cycle-regulated genes of the yeast Saccharomyces cerevisiae by microarray hybridization, Molecular Biology of the Cell, 9, pp. 3273–3297.
23. Tamayo, P., Slonim, D., Mesirov, J., Zhu, Q., Kitareewan, S., Dmitrovsky, E., Lander, E. S. and Golub, T. R. (1999). Interpreting patterns of gene expression with self-organizing maps: methods and application to hematopoietic differentiation, *Proc. of the National Academy of Sciences of the United States of America*, 96, pp. 2907–2912.
24. Tavazoie, S., Hughes, J. D., Campbell, M. J., Cho, R. J. and Church, G. M. (1999). Systematic determination of genetic network architecture, Nature Genetics, 22, pp. 281–285.
25. Walpole, R. E., Myers, R. H., Myers, S. L. and Ye, K. E. (2011) *Probability & Statistics for Engineers & Scientists*, (Pearson, USA).
26. Wen, X., Fuhrman, S., Michaels, G. S., Carr, D. B., Smith, S., Barker, J. L. and Somogyi, R. (1998). Large-scale time series gene expression mapping of central nervous system development, *Proc. National Academy of Sciences of the United States of America*, 95, pp. 334–339.
27. Xu, H. L., Liu, Y. H. and Wand, S. T. (2008). Autoregressive-model based dynamic fuzzy clustering for time-course gene expression data, Biotechnology, 7, pp. 59–65.
28. Yeung, K. Y., Haynor, D. R. and Ruzzo, W. L. (2001). Validating clustering for gene expression data, Bioinformatics, 17, pp. 309–318.
29. Yeung, K. Y. and Ruzzo, W. L. (2001). An empirical study of principal component analysis for clustering gene expression data, Bioinformatics, 17, pp. 763–774.
30. Zhao, W., Serpedin, E. and Dougherty, E. R. (2009). Spectral preprocessing for clustering time-series gene expressions, EURASIP Journal on Bioinformatics and System Biology: Article ID 713248, 10 pages.
31. Zhou, X., Wang, X., Dougherty, E. R., Russ, D. and Suh, E. (2004). Gene clustering based on cluster-wide mutual information, Journal of Computational Biology, 11, pp. 147–161.

Chapter 2

Gene ontology-based analysis of time series gene expression data using support vector machines[a]

Pei-Lin Chen, Rong-Ming Chen* and Been-Chian Chien
Department of Computer Science and Information Engineering, National University of Tainan, Taiwan

Rouh-Mei Hu and Jeffrey J. P. Tsai
Department of Bioinformatics and Medical Engineering, Asia University, Taiwan

2.1 Introduction

This section introduces the background, motivation and goal of this study, and the structure of this chapter.

Organizing biological gene expression data has become critical with the vigorous development of biotechnology and a continued increase in biological gene expression data. Bioinformatics was established to respond to the massive amount of gene data, which can no longer be analyzed manually. Bioinformatics combines various information technologies to enable faster calculations and classifications of voluminous gene data and provides them to biologists for analyses and interpretations.

At the biological level, correlations among genes are typically expressed as their similarities in time series gene expression data or time

[a] An earlier version of this study in Chinese was presented at The 2013 National Computer Symposium (Domestic Poster Track), Taiwan, Dec. 13-14, 2013.
*Corresponding author.

displacement. From the biological perspective, genes with similar patterns of time series gene expression data may be regarded as coregulated genes; these genes are typically applied to construct gene regulatory networks in systems biology [5]. Clustering has been performed in several studies to analyze genes with similarities in time series gene expression data [4, 32, 39, 40, 43, 45]. Genes with similar time series gene expression data have been clustered into the same groups through unsupervised clustering. Whether these genes truly share similar biological implications and functions requires further examination. Simply expanding the effectiveness of clustering algorithms may not guarantee that genes clustered into the same groups share similar biological implications. The results of unsupervised time series gene expression data clustering have been assessed using the silhouette index [48] or the adjusted Rand index [63]. The most favorable clustering result among previous studies only scored approximately 50–75%; an inaccuracy of up to 25–50% was still identified [4, 20, 21, 40]. Such an inaccuracy may have caused a significant proportion of genes in the same groups to differ in their biological implications and functions.

To improve clustering accuracy, some scholars have grouped time series gene expression data through supervised clustering [9, 47]. Supervised clustering involves training a clustering system by using sample data and establishing classification models before clustering. Therefore, if the gene expression data classified according to the functional classes of genes are used to train a clustering system and establish models, then the accuracy of supervised clustering may surpass that of unsupervised clustering in gene classifications. Studies that have incorporated supervised clustering to analyze gene expression data are discussed in the next section.

Studies have also applied information from gene ontology (GO) [2, 8] to effectively analyze functional genomics, proteomics, and microarray gene expression data [3, 18, 41, 54, 59]. For example, Tseng and Yu [54] organized microarray gene expression and GO classification data and created a multi-data gene scoring system to acquire informational genes. Conventional classification algorithms were then employed to test the system with the gene expression data of colon cancer and normal specimens. The result revealed that using GO classification data effectively improved the microarray gene expression data classification accuracy by 10–16%. In this study, support vector machines (SVMs) were corresponded with GO functional data to explore the supervised clustering of time series gene expression data.

Previous studies that have incorporated supervised clustering based on general gene function classifications have indicated that genes that are classified in the same functional classes may contain different GO terms, and the GO terms of genes with different functions may overlap with one another [2, 8, 17, 47]. Therefore, this study applied SVMs to explore GO-based supervised clustering for gene expression data and thereby improve the gene classification accuracy [17]. Three experiments were conducted to assess the effectiveness of the various clustering methods applied in this study. First, cross validation (CV) was conducted through the use of real gene expression data; the results were compared to those of *k*-means [33] and hierarchical methods [13], two widely used methods. Then, SVMs were incorporated to analyze the difference between conventional gene function classification-based and GO-based supervised clustering methods.

The remainder of this chapter is structured as follows. Section 2.2 presents the background knowledge and a literature review of time series gene expression data, gene function classifications, GO, and the current methods employed to process time series gene expression data. Section 2.3 describes the research methods, including data preprocessing, SVMs, the two clustering methods applied in the experiments, and class imbalance in data sampling. Section 2.4 explains the experimental design, which includes the content, goals, and parameter settings of three distinct experiments. Section 2.5 presents and discusses the effectiveness of the proposed method according to the results of the experiments, and compares it to those of methods applied in previous studies. Section 2.6 concludes this chapter and proposes suggestions for further studies.

2.2 Preliminaries

This section introduces the background knowledge relevant to this study and briefly describes the basic concepts and literature regarding various clustering methods. First, time series gene expression data are introduced as the primary targets of analysis in this study. Second, the two major types of clustering methods, supervised and unsupervised clustering, are described. Finally, gene functions and GO terms, which were used as the bases for gene clustering, are discussed.

2.2.1 *Time Series Gene Expression Data*

Microarray chips, alternatively referred to as gene chips, are characterized by their low production cost, fast data processing, and small physical volume–large quantity advantage. Therefore, numerous biologists have used microarray chips to examine gene expressions. Each microarray chip contains many small and concentrated probes, which display fluorescent reactions of varying levels of intensity through specific technical analyses. These reactions are converted to numerical data through image processing. The numerical data contain abundant genetic information for analyzing the significant differences among genes. The reaction values of each gene must be observed and recorded under various conditions or after a period of time; each period of time is not required to be equal to another [24].

2.2.2 *Gene Function Classification*

Gene functions are classified on the basis of the protein functions transcribed and translated from genetic information. The data on the functional classes of genes applied in this study were sourced from the Munich Information Center for Protein Sequences Yeast Genome Database (MIPS Yeast Genome Database, abbreviated as MYGD) [9, 44, 49]. The 227 genes selected from the MYGD were classified into six types of functions, namely tricarboxylic acid cycle (TCA), respiration (RESP), cytoplasmic ribosomes (RIBOs), proteasome (PROTEAS), histones (HISTs), and helix-turn-helix proteins (HTHs).

Scholars such as Brown *et al.* [9] have applied the six types of functions in experiments because the TCA, RESP, RIBO, HIST, and PROTEAS genes feature not only the same functions but also similar time series gene expression data patterns with other genes in their respective groups. HTHs were used as the control group because they exhibited less similar time series gene expression data patterns with one another. These genes could be divided into different groups through unsupervised clustering. This study incorporated a total of 227 time series gene expression data that contained all these six types of functions. The time series of two of the gene functions were illustrated in line charts to observe their patterns.

2.2.3 *Gene Ontology (GO)*

GO, a functional annotation framework for genes developed by the Gene Ontology Consortium [2, 8], is a hierarchical structure that consists of a DAG. Each GO node comprises a GO term, and each term contains a unique term name and a seven-digit GO identification number (ID) that starts with GO. For example, the GO ID for glucose transport is GO: 0015758. The relationship among GO terms is represented using "edges" connecting the nodes, or parents and children, in which children are more specialized than parents. Unlike conventional hierarchical structures that are generally more rigid, each child can be associated with more than one parent in GO. GO consists of three primary ontological structures, namely the biological process (BP), the molecular function (MF), and the cellular component (CC), all of which are the roots of their respective ontologies.

Unlike conventional functional class databases such as the well-known microprocessor without interlocked pipeline stages functional catalog [44, 49], GO is designed to provide the most detailed functional annotations on the protein as possible, such as the operational behaviors of gene products in cells. GO provides plentiful and detailed annotation data on protein functions, and the number of GO terms is greater than that of conventional functional class databases. Previous studies have maintained that an excessive number of terms might cause difficulties in processing annotations and inconsistencies in their designations [49]. However, comparing GO-based supervised clustering with conventional supervised clustering reveals that genes that are clustered in different function groups might contain the same GO terms, because the annotations of GO terms are more detailed. Therefore, this study hypothesizes that the accuracy of the GO-based gene classification method is superior to that of the conventional gene classification method in clustering time series gene expression data.

2.2.4 *Cluster Analysis of Time Series Gene Expression Data*

A cluster analysis involves analyzing, classifying, comparing, and thereby grouping genes with the same or similar expression patterns or biological implications. According to the learning modes involved, a cluster analysis is divided into supervised and unsupervised clustering. Most studies on gene expression data clusters have adopted unsupervised clustering, but some have employed supervised learning. Supervised and unsupervised clustering are briefly introduced in the following two sections.

2.2.4.1 *Supervised Clustering*

Two main types of supervised clustering methods have been adopted to analyze gene expression data. One type is based on probability distribution models [47] and exhibits a superior accuracy to that of unsupervised clustering methods and SVMs [9]. However, this method requires all gene expression values to be normally distributed. The other type of method involves using the known functional classes of genes for supervised clustering. Numerous studies have explored various supervised learning methods [35], such as SVMs [57, 58], Fisher's linear discriminant analysis [27], Parzen windows [7], C4.5 decision trees [46], and MOC1 decision trees [61]. SVMs feature considerable flexibility in selecting similarity functions, exhibit the characteristics of sparse solutions in processing voluminous sets of data, and enable solving problems involving nonlinear classification and high-dimensional feature spaces. Therefore, SVMs can be applied to gene expression data analyses [9, 11, 62]. Brown *et al.* [9] indicated that SVMs outperform the four other standard supervised learning methods in classifying gene expression data. Accordingly, SVMs were employed for supervised learning and establishing a classification system in this study.

2.2.4.2 *Unsupervised Clustering*

Unsupervised clustering is faster and less labor intensive than supervised clustering because it does not require manual data creation in the beginning. However, unsupervised clustering requires grouping gene data in advance. The clustering results are profoundly affected if no appropriate number of groups is set before clustering. Furthermore, unsupervised clustering is apt to affected by noise; if a massive number of outliers are generated in the gene expression data, the clustering results are substantially compromised. Therefore, the results of unsupervised clustering are less robust than those of supervised clustering. Unsupervised clustering is typically executed according to the distances among data, which are most commonly calculated as L_1-norm distances, L_2-norm distances, and cosine similarities. Some well-known unsupervised clustering approaches include *k*-means, hierarchical, self-organizing maps in artificial neural networks [34], and the adaptive resonance theory [12].

2.3 Methods

GO terms were used to examine the supervised clustering of time series gene expression data using SVMs. The procedures of the GO-based supervised clustering are listed as follows: (1) establish a classifier using the time series gene expression data of known GO terms as the training data; and (2) use the classifier to cluster the training data. Using the GO-based classifier for supervised clustering enables a gene to be simultaneously clustered into multiple types of functions; when the GO terms of a gene are identified, the functional classes of the genes can be predicted according to their corresponding GO terms [53]. In addition, when the GO terms of a gene do not correspond to any of the known genetic functions, it implies that the gene may have new functions. In this section, the gene expression data preprocessing and SVMs required for the clustering process are discussed, the GO-based supervised clustering method is presented, and the sampling approach to data with a class imbalance is also explained.

2.3.1 *Data Preprocessing*

Deviation, conversion, and noise may cause similar time series gene expression data to become different. These problems are averted during microarray gene expression tests and image data value conversion, if possible. In addition, missing values are the primary focus in data preprocessing when gene expression data are downloaded from the Internet. Solutions to missing values adopted by numerous studies include replacing them directly with zeroes, deleting the first *n* similar time series gene expression data at the temporal points outside the missing values and replacing the missing values with the mean value of these *n* time series at the temporal points of the missing values, and interpolation [19, 22]. Because the unknown errors caused by the existing missing value assessment methods must be eliminated first, in this study, before processing the missing values, the gene expression data downloaded from the yeast gene database were directly applied in the experiment to prevent potential errors in the subsequent result comparisons caused by the processing of missing values. In some studies, genes without significant changes in the expression values were removed [4]. Although such a method may improve the experimental result, this type of gene was initially included in the experiment of the present study.

2.3.2 *Support Vector Machines (SVMs)*

SVMs, which enable supervised learning, are a critical machine learning method that is based on statistical learning theory [10, 50, 55-57]. SVMs can process small-sample, nonlinear, and high-dimensional classification problems. SVMs have been broadly and successfully incorporated to solve classification problems involving biological information, document classification, and image identification [1, 25, 31, 37, 58]. SVMs involving diverse classifications can be divided into two types, one-against-one and one-against-rest, which feature their own advantages and disadvantages.

One-against-one involves a voting process, in which a set of gene data that contains n distinct functional classes can generate $n(n-1)/2$ different classifiers. Assuming that five different classes of functions (hereafter referred to as 1, 2, 3, 4, and 5) are incorporated into an experiment, pairing two of the classes generates an SVM classifier. Thus, a total of 10 distinct SVM classifiers are created: (1,2), (1,3), (1,4), (1,5), (2,3), (2,4), (2,5), (3,4), (3,5), and (4,5). When a test datum entering (1, 2) is clustered into 1, the frequency value of 1 is increased by one; otherwise, that of 2 is increased by one. Through the use of these 10 classifiers, the class that acquires the highest frequency value is the class that the test datum belongs to. Through the one-against-one approach, the number of data among the classes does not deviate excessively from one another, thus reducing the likelihood of a class imbalance in data sampling. However, when a gene belongs to more than one class, an uncertainty results in clustering. For example, when a test datum contains both Functions 1 and 2, clustering the datum in either 1 or 2 through (1, 2) impairs the quality in data sampling. Furthermore, when the number of classes of functions is excessively high, a massive number of classifiers must be created, costing valuable time.

One-against-rest classifications are simpler and much less time intensive than one-against-one. Through this method, K classifiers can be created out of K classes; test data are then clustered to their respective classes through these K classifiers. If there are four different classes, then the first SVM classifier will designate the first class as a positive class and the other three as negative classes, and so on. Eventually, four distinct classifiers are created. A test datum is then clustered through these four classifiers. If the datum is tested positive in the first classifier, then the datum contains the first class of functions. One-against-rest is time-efficient and convenient but may cause a class imbalance in data sampling.

This study adopted SVMs involving the one-against-rest method. The problem of class imbalance and its solution are detailed in Section 2.3.4 and the Experimental Design section.

2.3.3 *GO Term-Based Supervised Clustering*

Applying SVMs for classification involves two procedures. The first is to collect known GO term gene expression data and use them to train their corresponding SVM classifiers. The second procedure requires clustering genes through the use of these SVM classifiers. Tools incorporating SVMs include the famous Library for SVMs (LIBSVM) [14] and MATLAB.

Imbalance in the number of positive- and negative-class samples in an SVM experiment typically causes bias and errors in classification. Therefore, corrections must be performed at the beginning of the experiment through the adjustment of the Kernel matrix value during SVM optimization [9], data sampling, and boosting [6, 15, 16, 26, 29, 51, 60]. Undersampling, generally regarded as an effective correction method, was adopted in this study to solve the imbalance in the number of training samples.

Figure 2.1 illustrates the processes of the GO-based SVM time series gene expression data cluster analysis. Time series gene expression data with the same GO terms were selected as the positive training sample, and an equal number of time series with different GO terms were selected as the negative training sample. These samples were labeled according to their respective GO terms for the SVM model training. For an *L* number of different GO terms, *L* groups of training samples and labels are generated, and *L* SVM classifiers are created. The same test samples were categorized using *L* different classifiers to examine and calculate the statistical accuracy of the classifiers in categorizing the GO terms of the samples [42].

2.3.4 *Class Imbalance in Data Sampling*

Class imbalance problems (CIPs) are encountered when training samples are selected through conventional gene-function or GO-based classifications and affect the final classification results. The genes in test data can be categorized into classes that have greater numbers of genes than others, even though the genes do not belong the class.

Assume that in a CIP involving binary classification, the number of genes classified in the minority of positive classes is P, the number of genes classified in the majority of negative classes is N, and N/P is considerably larger than one. This was the CIP encountered in this study. When N and P were left unadjusted, a majority of the test samples were classified in the negative classes. Therefore, CIPs are problems that cannot be neglected.

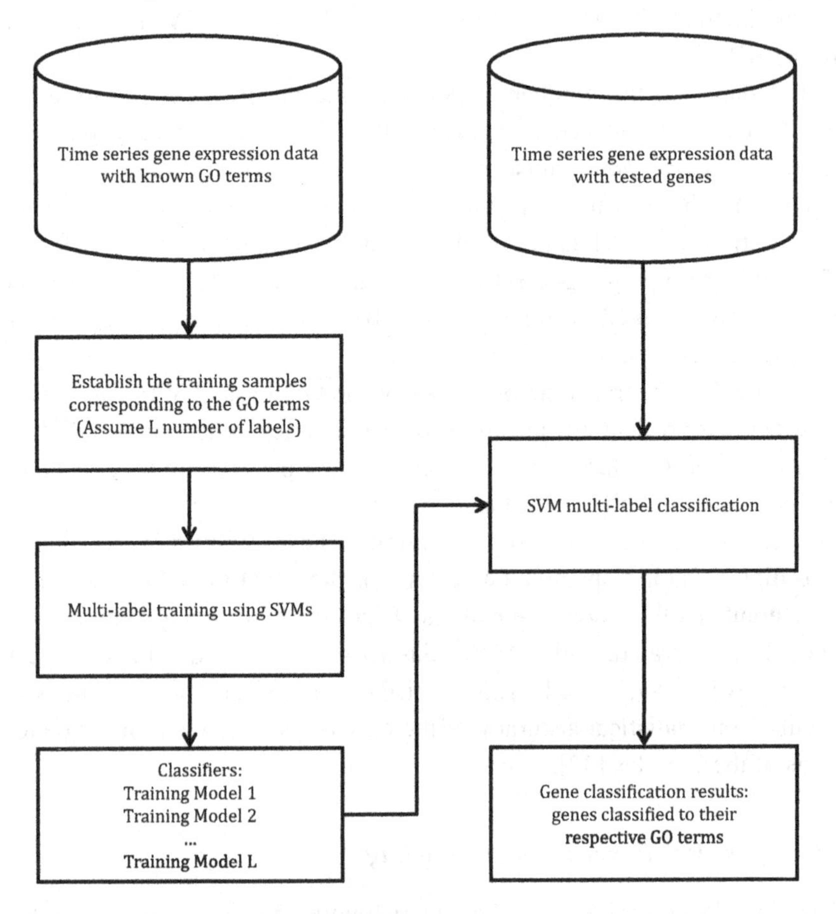

Fig. 2.1. Flow chart of GO-based SVM time series gene expression data cluster analysis.

Previous studies have addressed various solutions to CIPs, such as adjusting specific SVM parameters [9] and changing the number of samples according to *P* and *N* [38]. Adjusting specific SVM parameters requires experience, and parameter settings vary with the types of data. The number of samples can be changed through over- or undersampling

[15, 36]; however, there are no theories to verify whether over-or undersampling provides the greater sampling accuracy. Therefore, both these methods have been widely applied in studies [6]. In the present study, all the genes belonging to a specific GO term were designated as the positive class, and those not belonging to the GO term were regarded as the negative class, thereby rendering the number of negative genes greater than that of positive genes. Undersampling was executed to select positive and negative genes in equal numbers. Because a gene may belong to more than one functional class, when the negative genes were randomly sampled, those that overlapped with the positive class were excluded in advance. In other words, in each group of training samples, genes that belonged to both the positive and negative classes were precluded to ensure accuracy of the classification result.

2.4 Experimental Design

This section describes the data, goals, and methods of the three experiments conducted in this study. The ideas in each of the three experimental designs are clarified as follows. The first experiment involved comparing the effectiveness of supervised clustering with that of unsupervised clustering. The second and third experiments differed in the characteristics of their training samples; the results of the three experiments are explained in the Section 2.5. Following, the CV process is introduced before presenting the experimental designs.

2.4.1 *Cross Validation (CV)*

K-fold CV, an approach to improve sample models, involves randomly dividing a group of samples into K subgroups. One subgroup is used as the test sample, and the other $K - 1$ subgroups are designated as training samples. The process is repeated K times. For example, if K is assumed to be 10; a group of sampled data is evenly divided into 10 subgroups, labeled with numbers. In the first validation, the first subgroup is used as the test sample, and the other nine subgroups are designated as training samples. In the second validation, the second subgroup is used as the test sample, and the other nine subgroups are designated as training samples, and so on. Thus, a 10-fold CV is executed. Finally, the accuracies of the 10 results are averaged to obtain the effective value of the experiment. Generally, the number of subgroups is determined according to the number of samples obtained for an experiment.

2.4.2 *Experimental Design (1)*

In this experiment, SVMs and the conventional *k*-means and hierarchical unsupervised clustering methods were adopted to cluster the time series gene expression data. The data, goal, method, and parameter settings are detailed in this section, and the effectiveness and result comparison are thoroughly described in Section 2.5.

2.4.2.1 *Experimental Data*

This subsection introduces the gene expression and GO term data used in this experiment, as well as the reason and method for acquiring the data, thereby clarifying the data content of this experiment.

This experiment incorporated two types of gene expression data files. The first was gene association files (GAFs), which contained GO IDs, GO classes, and gene names, and could be downloaded from the GO website. The other was a gene expression data file. In this experiment, the gene expression data of *Saccharomyces cerevisiae* were employed [23]. The dataset comprised 6,149 gene expression data. Each time series gene expression data featured 17 time points. This study used gene expression data downloaded from the Saccharomyces Genome Database [28] because they have been incorporated as experimental data in numerous studies [47, 52]. In addition, the proportion of missing values in this data was only 0.08%, indicating that the missing values did not affect the result significantly.

As for the GO term data, typically, more than one gene is annotated with a specific GO ID. However, the number of corresponding genes for each GO ID varies. An insufficient number of training samples would affect the representativeness of the SVM models; therefore, at the beginning of this experiment, GO ID data with the numbers of their corresponding genes exceeding the threshold of 30 were selected from GAFs. Among the three GO classes (BP, MF, and CC), BP was adopted in this experiment; if either of the other two classes were selected, the method would remain the same. Subsequently, the time series gene expression data corresponding to each gene were selected from the gene expression data files. The resulting data files were then used as the experimental data for this study. A total of 10 distinct GO ID pairs were selected in this experiment. To equalize the number of genes in each GO ID pair, 19 GO IDs with approximate numbers of genes were selected from the filtered GAFs. These were GO:0006974, GO:0051301, GO:0006629, GO:0045944, GO:0006950, GO:0055085, GO:0002181, GO:0006281, GO:0006468, GO:0042254, GO:0006412, GO:0007049, GO:0006364, GO:0016310, GO:0006397, GO:0006355, GO:0015031, GO:0008152, and GO:0055114.

2.4.2.2 *Content of the Experiment*

This subsection briefly explains the core content of this experiment: the goal, method, and parameter settings of the experiment.

The *k*-means and hierarchical methods, two conventional unsupervised clustering methods, are based on the time series expression distance among genes. Ideally, genes clustered into one-group share similar or same functions. However, many genes that are close to one another rarely share the same genetic functions. Therefore, in this experiment, the samples were classified according to their GO IDs through the SVMs and the two conventional unsupervised clustering methods, and the cluster results were compared to verify whether the classification accuracy of the SVM algorithm was superior to that of the *k*-means and hierarchical methods.

To verify the effectiveness of the SVMs, a total of 10 GO ID pairs were selected. Genes that contained the GO IDs of both the positive and negative classes were precluded, and the genes corresponding to each GO ID were labeled as positive or negative classes. Considering the number of collected data and the balance between positive and negative classes on the numbers of samples in each GO ID pair, 105 genes were randomly selected from each of the two classes, for a five-fold CV. Each data group was divided into five subgroups; four were designated as training samples, and the last was used as the test sample for verifying the cluster results.

MATLAB was employed for three types of clustering algorithms. *k*-means clustering was performed using the default MATLAB parameters, and hierarchical clustering was conducted using the hierarchical agglomerative algorithm. The distances among the genes were calculated in Euclidean distances, and the distances among the clustered genes were calculated in average linkages. Because GO ID pairs were used to verify the effectiveness of the algorithms used in this experiment, the preliminary number of clusters for the *k*-means and hierarchical methods was set as two.

SVM algorithms require setting up more parameters than in the *k*-means and hierarchical methods. Gaussian radial basis function was selected for *kernel_function*, the primary parameters of which are *rbf_sigma* and *penalty parameter*. After numerous tests, three types of *rbf_sigma* setting values (0.5, 1, and 100) were used, and two types of *penalty parameter* setting values (0.8 and 128) were incorporated.

2.4.3 *Experimental Design (2)*

The second experiment used MIPS functional classes as the classification criteria for the SVM test [44, 49] and differed from the first experiment. Therefore, this subsection reintroduces the data, goal, method, and parameter settings for this experiment. To obtain the training and test samples, two distinct methods were employed and are detailed in the remainder of this section. Because the result of this experiment was compared to that of the third experiment, only the experimental design is described in this section; the result is detailed and evaluated in Section 2.5.

2.4.3.1 *Experimental Data*

This experiment required two types of data: gene expression data and the genes selected from the MYGD.

Similar to the first experiment, *Saccharomyces cerevisiae* gene expression data were employed in this experiment. A total of 6,239 gene expression data were used; each time series gene expression data contained 17 time points. The genes that did not belong to the MIPS functional catalogue were eliminated from the gene expression data; the remaining genes and their expression data were used in the experiment.

The second type of data, the functional class, used in this experiment were derived from the MYGD [9,44, 49]. In particular, 227 of the genes in this experiment belonged to TCA, HIST, HTH, PRO, RESP, and RIBO, and the others did not. Because *Saccharomyces cerevisiae* gene expression data lacked the time expression values for the YDL148Cgene, this gene was eliminated; the remaining 226 genes were incorporated as experimental data. Among the 226 genes, 17 belonged to TCA, 11 belonged to HIST, 16 belonged to HTH, 35 belonged to PROTEAS, 30 belonged to RESP, and 120 belonged to RIBO. Three of the genes, YDR178W,YKL148C, and YLL041C, belonged to both TCA and RESP.

2.4.3.2 *Content of the Experiment*

This subsection explains the design of this experiment, including the goal, method, and parameter settings. The MIPS functional classes were used as the classification criteria. To analyze the effect of the number of samples on the clustering result, two different methods for selecting the training and test samples were applied. These two methods only differed from each other in their sample selection approaches; they shared the same SVM parameter settings.

Through the use of the MIPS functional classes as the classification criteria, the goal of this experiment was to perform SVM supervised clustering on the time series gene expression data and evaluate the classification accuracy of the SVM. A higher classification accuracy enables SVMs to more accurately classify genes with unknown functions to their respective functional classes.

Two different training and test sample selection methods were incorporated in this experiment. One of the methods involved selecting training and test samples from the 226 genes that belonged to the aforementioned six functional classes and running a CV, thus requiring relatively few training samples. For example, the division of the 17 genes in the TCA class into a three-fold CV is set as CV1(5), CV2(6), and CV3(6), where the numbers in parentheses indicate the number of genes in each of the three TCA subgroups.

One of the subgroups was designated as the test sample, and the other two were used as training samples. For example, the five genes in CV1 were used as the test sample, and those of CV2 and CV3 were employed as training samples; a total of 12 genes were classified in the positive class. An equivalent number of negative genes among the HIST, HTH, PROTEAS, RESP, and RIBO classes were selected, with the total number equal to the number of positive genes. Thus, the training sample in this experiment consisted of 12 positive genes and 12 negative genes.

The selection of the test sample differed from that of the training samples in that it involved selecting an equal number of genes for the negative class to that of the genes for the positive class from each of the five functional classes. In the aforementioned example, five genes belonged to the positive class; thus, five genes were selected from each of the five functional classes to form the negative class. A total of 30 test samples were used in the example. Notably, the genes in the training samples must not overlap with those of the test sample. When the number of genes in a specific class was insufficient, an adequate number of genes from the other functional classes were selected. Specifically, the number of genes in RIBO exceeded five times the total number of genes from all the other functional classes that were not selected for the training samples. Therefore, the RIBO test sample genes were applied together with the genes from all the other functional classes that were not selected as training samples, as the test sample in this study.

The other selection method involved all 226 genes in the six functional classes, but did not undergo CV. For example, all 17 TCA genes were used as positive genes, and 17 random genes that did not overlap with these genes were selected from the gene expression data of Saccharomyces cerevisiaeas negative genes. Thus, a total of 34 genes were used for the training samples.

Similar to the first experiment, this experiment used MATLAB to execute SVM algorithms. Therefore, the SVM parameters were the same as those in the first experiment; nine sets of experimental parameters were established. The first selected sample in this experiment required one more parameter. Because some of the functional classes contained 10–20 genes each, excessively dividing the data in CV would cause insufficiency in the training and test samples and affect the experimental result. Therefore, a three-fold CV was conducted in this experiment.

2.4.4 *Experimental Design (3)*

Similar to the second experiments, this experiment incorporated the SVM algorithm for supervised learning. However, the classification indices in this experiment were GO IDs. Therefore, this subsection reintroduces the goal, method, and parameter settings in this experiment; the result is compared to that of the second experiment in Section 2.5.

2.4.4.1 *Experimental Data*

This experiment required two types of data: the GAFs as applied in the first experiment, and the genes of the six functional classes from the MYGD in the second experiment.

Numerous GO term data were employed in this experiment. From the GAFs, the GO ID data with the numbers of their corresponding genes exceeding the experiment threshold of 30 were selected. Among the three types of GO classes, the BP was selected for this experiment. Subsequently, the expression time series corresponding to each gene were selected from the gene expression data files. A total of 185 GO IDs were incorporated, and the number of genes for each ID varied; some GO IDs had only one corresponding gene each, whereas others exhibited more than 400.

Similar to the first experiment, this experiment employed *Saccharomyces cerevisiae* gene expression data. A total of 6,239 gene expression data were used; each time series gene expression data exhibited 17 time points. From this data, the 185 GO IDs of the 226 genes were selected. The genes that did not contain these 185 GO IDs were then eliminated from the 6,239 gene expression data.

2.4.4.2 *Content of the Experiment*

This subsection introduces the goal, method, and parameter settings for this final experiment of the study. The result is presented in Section 2.5.

The MIPS functional classes are broader classification classes than GO IDs. Analyzing the gene data of the six classes from the MYGD revealed that the genes clustered in the same functional classes did not necessarily share the same GO IDs. Therefore, using the GO IDs as the classification indices, this experiment verified whether genes with the same GO IDs could be clustered in the same functional classes, thereby effectively improving the SVM time series gene expression data classification accuracy.

This experiment required two types of files. One was the 227-gene file downloaded from the MYGD. Similar to the second experiment, YDL184C was removed and the remaining 226 genes were used. The other type was the GAFs that can be downloaded from the GO website. A total of 185 GO IDs corresponding to the 226 genes were identified from the files, and all the genes corresponding to these 185 GO IDs were selected from the GAFs.

The 185 GO IDs were used to create 185 SVM classifiers. The genes corresponding to each of the GO IDs were grouped as the positive genes. Subsequently, an equal number of genes not corresponding to the GO IDs (thus avoiding overlap with the positive genes) were then randomly selected from the processed files as the negative genes. Thus, a total of 185 GO ID pairs were established. For example, in the GAFs, the 38 genes corresponding to GO: 0000002 were used as the positive genes in the training samples. From the processed GAFs, 38 random genes not corresponding to GO: 0000002 were then selected as the negative genes. Each GO ID enabled the creation of an SVM classifier.

The 226 genes downloaded from the MYGD were used as the test sample, and their time series gene expression data were applied in the SVM classifier. If the classification result of one of the genes was positive, then the gene contained the GO ID of the classifier; otherwise, the gene did not contain this GO ID.

Similar to the first two experiments, this experiment used MATLAB to implement its SVM algorithms. Therefore, the SVM parameters used in this experiment were the same as those in the second experiment, with a total of nine sets of parameters. Furthermore, because some of the GO IDs contained few or only one gene connotation, no CV was performed in this experiment; all the genes corresponding to the GO IDs were incorporated as training samples.

2.5 Results and Discussion

This section analyzes and compares the results of the three experiments. The first subsection introduces the approach for assessing the classification results. The second subsection presents a comparison between the clustering results of the GO ID-based SVM algorithm and the two unsupervised clustering methods detailed in Section 2.4.2 (the first experiment). The third subsection explains the comparison between the time series gene expression data clustering results of the MIPS-based SVM algorithm and the GO ID-based approach as described in Sections 2.4.3 and 2.4.4 (the second and third experiments). The final subsection presents an analysis of the effect of the DAG levels of GO IDs on the classification accuracy of the third experiment.

2.5.1 *Classification Result Assessment*

The functional accuracies of the classification results in this study were used to examine the effectiveness of the methods applied in the experiments, thereby assessing the probability of genes that share the same GO terms or functional classes to be accurately clustered into the same groups. Accuracy is commonly applied as a statistical index in a binary classification assessment. The effectiveness of classification is determined by whether a classifier accurately differentiates the positive and negative groups in test samples. In clustering the test samples, the classification of each gene may be generalized into one of the following four conditions: true positive (TP), true negative (TN), false positive (FP), and false negative (FN). The accuracy is defined as follows:

$$Accuracy = \frac{TP + TN}{TP + TN + FP + FN} \tag{2.1}$$

where the rate is represented as a percentage ranging from zero to one. The closer to one the value is, the higher the classification accuracy it indicates.

2.5.2 *Comparison of the GO ID-Based SVM Algorithm with Conventional Unsupervised Clustering Methods*

In the first experiment, all the GO ID pairs underwent the five-fold CV. The average clustering accuracies of the proposed, *k*-means, and hierarchical methods in the five-fold CV of the 10 GO ID pairs are visually presented in Fig. 2.2, which reveals that the classification accuracy of the GO ID-based SVM algorithm method significantly surpassed that of the two conventional unsupervised clustering methods.

2.5.3 *Comparison of the MIPS-Based and GO-Based SVM Algorithms*

This subsection presents a comparison of the results of the second and third experiments to verify whether the clustering accuracy of the GO-based SVM algorithm was superior to that of the MIPS-based SVM algorithms.

Table 2.1 presents the results of the three-fold CV that involved the MIPS-based SVM algorithm from the first sampling method, in which the accuracy of RIBO was the highest (95.5%). This may be because the number of RIBO genes was 120. In the three-fold CV test, the total number of training samples in each CV was 160; a higher number of training samples led to a more favorable classification result. The accuracy of the TCA was the lowest (69.6%), even though the TCA did not have the fewest genes among the six functional classes. HIST had the fewest genes (11) and a maximum of only 16 training samples in a CV, yet its clustering accuracy was as high as 87.9%. The average accuracy for the six classes was 82.7%. Table 2.2 presents the results of the three-fold CV that involved the MIPS-based SVM algorithm from the second sampling method, in which RIBO also exhibited the highest accuracy. This was because among the 226 test samples, 120 were clustered into the positive class. The TCA had the poorest result, with an accuracy of only 65%. The average accuracy was 81.1%, slightly lower than that of the first MIPS-based algorithm; however, this difference was not significant. These results indicate that an insufficiency in training samples might have affected the clustering accuracy. Overall, the average accuracy of the MIPS-based SVM algorithms was about 82%.

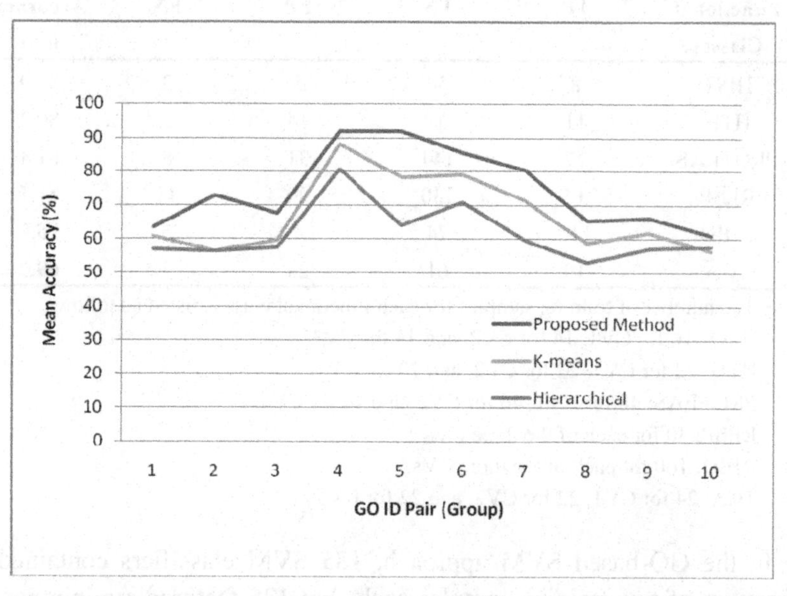

Fig. 2.2. Comparison of the average clustering accuracies of the proposed, k-means, and hierarchical methods in the five-fold CV of the 10 GO ID pairs. (Some of the data are also presented in an unpublished project report for the Ministry of Science and Technology, Taiwan, which sponsored this study. Project Number: NSC 101-2221-E-024-024).

The results of the GO-based sample selection method are discussed as follows. A total of 185 GO-based training models were adopted in this experiment. The number of training samples for each model varied substantially; some contained only two samples and others exhibited as many as 400. Therefore, the classification accuracies of the 185 classifiers were assessed using the total number of training samples. Table 2.3 presents the results of the third experiment. A higher number of training samples yielded a higher average classification accuracy; when the number of training samples exceeded 30, the average classification accuracy exceeded 90%.

Table 2.1. Classification accuracy of the six MIPS-based functional classes from the first sampling method in the three-fold CV (%).

Functional Classes	TP	TN	FP	FN	Accuracy (%)
HIST	8	50	5	3	87.9
HTH	11	66	14	5	80.2
PROTEAS	27	144	31	8	81.4
RESP	17	130	20	13	81.7
RIBO	115	74	4	5	95.5
TCA	10	61	24	7	69.6

Note: The number of training samples for each functional class is listed as follows.
HIST: 16 for CV1, 14 for CV2, and 14 for CV3.
HTH: 22 for CV1, 22 for CV2, and 20 for CV3.
PROTEAS: 48 for CV1, 46 for CV2, and 46 for CV3.
RESP: 40 for each of the three CVs.
RIBO: 160 for each of the three CVs.
TCA: 24 for CV1, 22 for CV2, and 22 for CV3.

In the GO-based SVM approach, 185 SVM classifiers contained a minimum of two training samples each, and 125 featured a minimum of 10 training samples each. A total of 60 SVM classifiers (approximately a third of the total number of classifiers in this experiment) exhibited less than 10 training samples each. However, the average accuracy in this experiment (84.7%) was still higher than those of the MIPS-based SVM algorithms (82.7% for the first sampling method; 81.1% for the second sampling method).

The results revealed that the clustering result of the GO-based SVM algorithm was superior to those of the MIPS-based algorithms. When the number of training samples exceeded 40, the average classification accuracy was improved to as high as 93%. Following the continuing expansion of and updates to the GO database, the number of genes corresponding to the GO IDs will continue to increase, thereby promoting the practicality of the proposed method.

Table 2.2. Classification accuracy of the six MIPS-based functional classes from the second sampling method in the three-fold CV (%).

Functional Classes	Number of Training Samples	TP	TN	FP	FN	Accuracy(%)
HIST	22	11	148	67	0	70.4
HTH	32	16	173	37	0	83.6
PROTEAS	70	35	163	28	0	87.6
RESP	60	30	160	36	0	84.1
RIBO	240	120	97	9	0	96
TCA	34	17	130	79	0	65

Table 2.3. Average classification accuracies of all GO ID-based SVM classifiers corresponding to the range of the number of training samples (%).

Number of Training Samples	Corresponding Number of SVM Classifiers	Average Accuracy (%)
≧2	185	84.6
≧10	125	87.1
≧20	91	89.4
≧30	74	91.6
≧40	58	93.3
≧50	53	92.9
≧60	47	93.2
≧70	37	95.3
≧80	34	95.5
≧90	31	96
≧100	28	95.8

2.6 Conclusions and Future Research

The conclusions and future research are discussed in two separate subsections. In the conclusion subsection, the problems encountered in the experiments are summarized, and the implications of the experimental results are reviewed. In the future research section, the viable directions and potential problems for further studies are discussed.

2.6.1 *Conclusions*

This study applied a GO-based SVM algorithm to explore the unsupervised clustering of time series gene expression data. Genes that share identical expression time series patterns may not have the same or similar functions and biological implications. Therefore, gene expression data with the same GO terms were used for training SVMs to classify genes with the same GO terms into the same groups.

Three experiments were conducted in this study, and their results were assessed using two approaches. The first experiment involved comparing the GO-based SVM algorithm with two conventional unsupervised clustering methods, and revealed that the SVM algorithm was significantly more effective in clustering genes with the same or similar biological implications into the same groups. The second and third experiments, both of which incorporated SVMs, differed in their training sample selection criteria. The second experiment employed MIPS as its criteria, whereas the third experiment employed GO terms. The results revealed that the proposed method in this study enabled more accurate gene classification than did the MIPS-based approach.

Finally, the correlation between the DAG levels of GO terms and the corresponding SVM classification accuracies were examined. The experimental results indicated that when the number of training samples exceeded a certain threshold, the high and stable SVM classification accuracies of most of the GO terms on Levels 5 and 6 were noted. Because DAGs are vital information in GO, further studies can focus on the correlation among the number of training samples, the DAG levels of GO terms, and their corresponding SVM classification accuracies.

2.6.2 *Future Research*

The problems encountered in these experiments and viable directions for further studies are listed as follows:

(1) Although the use of microarray chips to obtain time series gene expression data is easy and well developed, missing values remain a critical problem. Numerous experiments have been conducted to solve the missing-value problem. Future studies can incorporate various missing-value solution approaches to improve classification accuracy.

(2) This study employed the time series gene expression data for Saccharomyces cerevisiae [23], which have been widely employed as experimental data in previous studies. Other widely known types of gene expression data can also be used in future studies.

(3) Following continuing updates of the GO database, experiments can be regularly conducted to analyze the changes in gene clustering results and verify the feasibility of the proposed method in this study.

(4) As mentioned in this study, class-imbalance in data sampling is a problem encountered in the one-against-rest method. In randomly selecting negative genes in an equal number to that of positive genes, informational genes may be overlooked and deviant ones may be included in the sample. Because gene selections profoundly affect the clustering and classification results, an improved solution to class imbalances in data sampling would increase the classification accuracy of the proposed method.

(5) The data in this study were randomly divided into subgroups through CV. These subgroups were then used to verify the clustering result of the proposed method. A three-fold CV test was performed using one subgroup as the test sample and the other two as training samples. Thus, the test was conducted three times, and the test data were averaged to establish the clustering results. In future studies, the frequency of the CV can be increased to verify whether a significant difference exists in the test results.

Applying the aforementioned suggestions and experimenting with different SVM parameter settings and functions can enable a higher accuracy in classifying genes with similar biological implications into the same groups.

Acknowledgements

This work was supported in part from the Ministry of Science and Technology, Taiwan [Grant numbers: NSC 101-2221-E-024-024, MOST 102-2221-E-024 -019, MOST 103-2221-E-024-014 and MOST 104-2221-E-024-018].

References

1. Abe, S. (2005). *Support Vector Machines for Pattern Classification* (Springer-Verlag, London, UK).
2. Ashburner, M., Ball, C. A., Blake, J. A., Botstein, D., Butler, H., Cherry, J. M., Davis, A. P., Dolinski, K., Dwight, S. S., Eppig, J. T., Harris, M. A., Hill, D. P., Issel-Tarver, L., Kasarskis, A., Lewis, S., Matese, J. C., Richardson, J. E., Ringwald, M., Rubin, G. M. and Sherlock, G. (2000). Gene ontology: Tool for the unification of biology, Nature Genetics, 25, pp. 25–29.
3. Azuaje, F., Al-Shahrour, F. and Dopazo, J. (2006). Ontology-driven approaches to analyzing data in functional genomics, Methods in Molecular Biology, 316, pp. 67–86.
4. Bandyopadhyay, S., Mukhopadhyay, A. and Maulik, U. (2007). An improved algorithm for clustering gene expression data, Bioinformatics, 23, pp. 2859–2865.
5. Balasubramaniyan, R., Hüllermeier, E., Weskamp, N. and Kämper, J. (2005). Clustering of gene expression data using a local shape-based similarity measure, Bioinformatics, 21, pp. 1069–1077.
6. Batista, G. E. A. P. A., Prati, R. C. and Monard, M. C. (2004). A study of the behavior of several methods for balancing machine learning training data, *Proc. ACM SIGKDD Explorations Newsletter*, pp. 20–29.
7. Bishop, C. (1995) *Neural Networks for Pattern Recognition* (Oxford University Press, New York, NY, USA).
8. Blake, J. A. and Harris, M. A. (2008). The Gene Ontology (GO) project: Structured vocabularies for molecular biology and their application to genome and expression analysis, Current Protocols in Bioinformatics, Chapter 7, Unit 7.2.
9. Brown, M. P. S., Grundy, W. N., Lin, D., Cristianini, N., C. Sugnet, W., Furey, T. S., Ares, M. and Haussler, D. (2000). Knowledge-based analysis of microarray gene expression data by using support vector machines, Proceedings of the National Academy of Sciences of the United States of America, 97, pp. 262–267.

10. Burges, C. J. C . (1998). A tutorial on support vector machines for pattern recognition. Data Mining and Knowledge Discovery, 2, pp.121–167.
11. Byvatov, E. and Schneider, G. (2003). Support vector machine applications in bioinformatics, Applied Bioinformatics, 2, pp. 67–77.
12. Carpenter, G. A. and Grossberg, S. (1987). A massively parallel architecture for a self-organizing neural pattern recognition machine, Computer Vision, Graphics, and Image Processing, 37, pp. 54–115.
13. Carr, D. B., Somogyi, R. and Michaels, G. (1997). Templates for looking at gene expression clustering, Statistical Computing & Statistical Graphics Newsletter, 8, pp. 20–29.
14. Chang, C. C. and Lin, C. J. (2011). LIBSVM: A library for support vector machines, ACM Transactions on Intelligent Systems and Technology, 2 pp. 1–27.
15. Chawla, N. V., Hall, L. O., Bowyer, K.W. and Kegelmeyer, W. P. (2002). Smote: Synthetic minority over-sampling technique, Journal of Artificial Intelligence Research, 16, pp. 321–357.
16. Chawla, N. V., Lazarevic, A., Hall, L.O. and Bowyer, K. W. (2003). SMOTEBoost: Improving prediction of the minority class in boosting, Principles and Practice of Knowledge Discovery in Databases, pp. 107–119.
17. Chen, P. L., Chen, R. M., Li, C. Y., Hu, R. M., Chien, B. C. and Tsai, J. P. (2012).Towards improving the biological relevance of time-series gene expression clustering using support vector machine, *Presented at the 23rd International Conference on Genome Informatics*, GIW (Poster).
18. Cheng, J. J, Cline, M., Martin, J., Finkelstein, D., Awad, T., Kulp, D. and Siani-Rose, M. A. (2004). A knowledge-based clustering algorithm driven by Gene Ontology, Journal of Biopharmaceutical Statistics, 14, pp. 687–700.
19. Ching, W. K., Li, L., Tsing, N. K., Tai, C. W., Ng, T. W., Wong, A. S. and Cheng, K. W. (2010). A weighted local least squares imputation method for missing value estimation in microarray gene expression data, International Journal of Data Mining and Bioinformatics, 4, pp. 331–347.
20. Chiu, T. Y., Hsu, T. C. and Wang, J. S. (2010). AP-based consensus clustering for time series gene expression data, *Proc. 20th International Conference on Pattern Recognition*, pp. 2512–2515.
21. Chiu, T. Y., Hsu, T. C., Yen, C. C. and Wang, J. S. (2015). Interpolation based consensus clustering for gene expression time series, BMC Bioinformatics, 16, pp. 117–133.
22. Choong, M. K., Charbit, M. and Yan, H. (2009). Autoregressive- model-based missing value estimation for DNA microarray time series data, IEEE Transactions on Information Technology in Biomedicine, 13, pp. 131–137.
23. Cho, R. J., Campbell, M. J., Winzeler, E. A., Steinmetz, L., Conway, A., Wodicka, L. , Wolfsberg, T. G., Gabrielian, A. E., Landsman, D., Lockhart, D. J. and Davis, R. W. (1998). A genome-wide transcriptional analysis of the mitotic cell cycle, Molecular Cell, 2, pp. 65–73.

24. DeRisi, J., Penland, L., Brown, P. O., Bittner, M. L., Meltzer, P. S., Ray, M., Chen, Y., Su, Y. A. and Trent, J. M. (1996). Use of a cDNA microarray to analyze gene expression patterns in human cancer, Nature Genetics, 14, pp. 457–460.
25. Dor, K. C., Chambwe, N., Srdanovic, M. and Campagne, F. (2010). BDVal: Reproducible large-scale predictive model development and validation in high-throughput datasets, Bioinformatics, 26, pp. 2472–2473.
26. Drummond, C. and Holte, R. C. (2003). C4.5, class imbalance, and cost sensitivity: Why under-sampling beats over-sampling, *Proc. International Conference on Machine Learning: Workshop Learn from Imbalanced Data Sets II*, pp. 1–8.
27. Duda, R. O. and Hart, P. E. (1973). *Pattern Classification and Scene Analysis* (Wiley, New York, NY, USA).
28. Engel, S. R., Dietrich, F. S., Fisk, D. G., Binkley, G., Balakrishnan, R., Costanzo, M. C., Dwight, S. S., Hitz, B. C., Karra, K., Nash, R. S., Weng, S., Wong, E. D., Lloyd, P., Skrzypek, M. S., Miyasato, S. R., Simison, M. and Cherry, J. M. (2014). The reference genome sequence of Saccharomyces cerevisiae: Then and Now, G3 (Bethesda), 4, pp. 389–398.
29. Freund, Y. and Schapire, R. (1996). Experiments with a new boosting algorithm, Proc.13th International Conference Machine Learning, pp. 148–156.
30. Freund, Y. and Schapire, R. (1999). A short introduction to boosting, Journal of Japanese Society for Artificial Intelligence, 14, pp. 771–780.
31. Idicula-Thomas, S., Kulkarni, A. J., Kulkarni, B, D., Jayaraman, V. K. and Balaji, P. V. (2006). A support vector machine-based method for predicting the propensity of a protein to be soluble or to form inclusion body on overexpression in Escherichia coli., Bioinformatics, 22, pp. 278–284.
32. Jain, A. K. and Dubes, R. C. (1988). *Algorithms for Clustering Data* (Prentice-Hall, Englewood Cliffs, NJ).
33. Jain, A. K. (2010). Data clustering: 50 years beyond K-means, Pattern Recognition Letters, 31, pp. 651–666.
34. Kohonen, T. (1990). The self-organizing map, Proceedings of the IEEE, 78 pp. 1464–1480.
35. Kotsiantis, S. B. (2007). Supervised machine learning: A review of classification techniques, Informatica, 31, pp. 249–268.
36. Kubat, M. and Matwin, S. (1997). Addressing the curse of imbalanced training sets: One-sided selection, *Proc. 14th International Conference on Machine Learning*, pp. 179–186.
37. Kumar, R., Kulkarni, A. J., Jayaraman, V. K. and Kulkarni B. D. (2004). Symbolization assisted SVM classifier for noisy data, Pattern Recognition Letters, 25, pp. 495–504.
38. Lin, Z. Y., Hao, Z. F., Yang, X. W. and Liu, X. L. (2009). *Several SVM Ensemble Methods Integrated with Under-Sampling for Imbalanced Data Learning* (Springer, Berlin, Germany) pp. 536–544.

39. Li, C. Y. (2011). *Spectral Processing and Autoregressive Modeling for Unsupervised Clustering of Time-Series Gene Expressions* (Master Thesis, National University of Tainan, Taiwan) (in Chinese).
40. Li, C. Y., Chen, R. M., Chien, B. C., Hu, R. M. and Tsai, J. P. (2011). Spectral processing and autoregressive modeling for unsupervised clustering of time-series gene expressions, *Proc. 16th Conference on Technologies and Applications of Artificial Intelligence*, TAAI, pp. 64–71 (in Chinese).
41. Maetschke, S. R., Simonsen, M., Davis, M. J. and Ragan, M. A. (2012). Gene Ontology-driven inference of protein-protein interactions using inducers, Bioinformatics, 28, pp. 69–75.
42. Manning, C. D., Raghavan, P. and Schütze, H. (2008). *Introduction to Information Retrieval* (Cambridge University Press, UK).
43. Maulik, U. and Bandyopadhyay, S. (2002). Performance evaluation of some clustering algorithms and validity indices, IEEE Transactions on Pattern Analysis and Machine Intelligence, 24, pp. 1650–1654.
44. Mewes, H. W., Freshman, D., Güldener, U., Mannhaupt, G., Mater, K., Mokrejs, M., Morgenstern, B., Münsterkötter, M., Rudd, S. and Weil, B. (2002). MIPS: A database for genomes and protein sequences, Nucleic Acids Research, 30, pp. 31–34.
45. Qin, Z. S. (2006). Clustering microarray gene expression data using weighted Chinese restaurant process, Bioinformatics, 22, pp. 1988–1997.
46. Quinlan, J. R. (1993). *C4.5: Programs for Machine Learning* (Morgan Kaufmann, San Francisco, USA).
47. Qu, Y. and Xu. S. (2004). Supervised cluster analysis for microarray data based on multivariate Gaussian mixture, Bioinformatics, 20, pp. 1905–1913.
48. Rousseeuw, P. (1987). Silhouettes: A graphical aid to the interpretation and validation of cluster analysis, Journal of Computational and Applied Mathematics, 20, pp. 53–65.
49. Ruepp, A., Zollner, A., Maier, D., Albermann, K., Hani, J., Mokrejs, M., Tetko, I., Güldener, U., Mannhaupt, G., Münsterkötter, M., Mewes, H. W. (2004). The FunCat, a functional annotation scheme for systematic classification of proteins from whole genomes, Nucleic Acids Research, 32, pp. 5539–5545.
50. Scholkopf, C., Burges, J. C. and Smola, A. J. (1999). *Advances in Kernel Methods* (MIT Press Cambridge, MA, USA).
51. Seiffert, C., Khoshgoftaar, T. M., Hulse, J. Van and Napolitano, A. (2010). RUSBoost: A hybrid approach to alleviating class imbalance, IEEE Transactions on Systems, Man, and Cybernetics - Part A: Systems and Humans, 40, pp. 185–197.
52. Tamayo, P., Slonim, D., Mesirov, J., Zhu, Q., Kitareewan, S., Dmitrovsky, E., Lander, E. S. and Golub, T. R. (1999). Interpreting patterns of gene expression with self-organizing maps: methods and application to hematopoietic differentiation, Proceedings of the National Academy of Sciences of the United States of America, 96, pp. 2907–2912.

53. The Gene Ontology Consortium (2010). The Gene Ontology in 2010: Extensions and refinements, Nucleic Acids Research, 38, D331–D335.
54. Tseng, V. S. and Yu, H. H. (2011). Microarray data classification by multi-information based gene scoring integrated with gene ontology, International Journal of Data Mining and Bioinformatics, 5, pp. 402–416.
55. Vapnik, V. N. (1995). *The nature of Statistical Learning Theory* (Springer-Verlag, New York, NY, USA).
56. Vapnik, V. N. and Cortes, C. (1995). Support vector networks, Machine Learning, 20, pp. 273–297.
57. Vapnik, V. N. (1998). *Statistical Learning Theory* (Wiley, New York, NY, USA).
58. Wang, L. (2005). *Support Vector Machines: Theory and Applications* (Springer, New York, NY, USA).
59. Warita, K., Mitsuhashi, T., Tabuchi, Y., Ohta, K., Suzuki, S., Hoshi, N., Miki, T. and Takeuchi, Y. (2012). Microarray and gene ontology analyses reveal downregulation of DNA repair and apoptotic pathways in diethylstilbestrol-exposed testicular Leydig cells, The Journal of Toxicological Sciences, 37, pp. 287–295.
60. Weiss, G. M. (2004). Mining with rarity: A unifying framework, *Proc. ACM SIGKDD Explorations Newsletter*, pp. 7–19.
61. Wu, D., Bennett, K., Cristianini, N. and Shawe-Taylor, J. (1999). Large margin decision trees for induction and transduction, *Proc.16th International Conference on Machine Learning*, pp. 474–483.
62. Yang, Z. R. (2004). Biological applications of support vector machines, Brief Bioinformatics, 5 pp. 328–338.
63. Yeung, K. Y. and Ruzzo, W. L. (2001). Principal component analysis for clustering gene expression data, Bioinformatics, 17, pp. 763–774.

Chapter 3

A comparative review of graph-based ensemble clustering as transformation methods for microarray data classification

Tossapon Boongoen* and Natthakan Iam-On[†,‡]
School of Information Technology,
Mae Fah Luang University
Chiang Rai 57100, Thailand
**t.boongoen@gmail.com*
†natthakan@mfu.ac.th

Recently, the use of ensemble data matrix as a transformed space for classification has been put forward. Specific to the problem of predicting student dropout, the matrix generated as part of summarizing members in a cluster ensemble is investigated with a number of conventional classification methods. Despite the reported success in comparison to the case of original data and other attribute reduction techniques like PCA and KPCA, the study is limited to only one ensemble matrix that is created by the link-based ensemble approach or LCE. To provide an comparative review with respect to the aforementioned problem, this paper includes the experiments and findings obtained from the use of different graph-based ensemble algorithms as data transformation methods for microarray data classification. The empirical study can be hugely useful particularly for those working in bioinformatics, and generally applicable to any classification problem. Besides, the review initiates another interesting challenge for many researchers in the field of cluster ensemble to coupling their models with this hybrid, clustering-classification learning.

Keywords: Cluster ensemble, data matrix, transformation method, classification, gene expression.

‡Corresponding author.

3.1 Introduction

The ability to identify biomarkers that imply a particular biological state, has received a great deal of attention among those working in biomedical and health fields. This recently emerges as one of the most important topics in cancer diagnostics. With established technology like microarray and mass spectrometry, it is possible to conduct a quantitative investigation on discriminating diseased samples from the collection of control ones.[1] Specific to the innovation of microarray, bioinformatics and medical researches have been lifted to another level, where a tool is made available for the examination of gene expression profiles. This has inspired novel applications such as identification of distinctly expressed genes for following molecular studies or the determination of drug therapy response,[40,44,45] and the development of classification systems for decision-support diagnosis.[8,42]

The problem of disease class prediction has gained a momentum with respect to the birth of DNA microarrays, especially for cancer cases. The principal objective of such a task is to categorize a sample under question to one of the pre-defined diagnostic groups or classes, based on the pattern exhibiting by its gene expression profile.[47] A dilemma encountered with an expression data matrix is the fact that a number of samples is frequently much smaller than that of attributes or genes. Yet, in the context of classification, not all attributes are informative, even a small collection of genes is truly relevant.[32,46] In response, several techniques have been devised to select a subset of genes to build an accurate classifier. These include univariate methods, where each gene is separately considered for its correlation with the target classes. Initial attempts are the Wilcoxon test,[10] the partial least squares (PLS)[34] and the mixture model algorithm.[37] The other family that is referred to as multivariate selection investigates the dependency among genes during the process of seeking the desired gene subset.[4] Examples of these are the Bayesian stochastic search variable selection method[18] and the multivariate Bayesian model.[30]

In addition to the selection approach, a set of so-called dimensionality reduction techniques has been promoted ans successfully exploited for data classification.[3] Generally, these transform the original data attributes to a new, smaller, more meaningful set for analysis and visualization. One the well-known algorithm, Principal component analysis (PCA),[27] has been employed to reduce the dimension of gene expression data, which allows comprehension of similarities among the biological samples and identification of noise. Despite its recognition, PCA may fall short from expectation sometimes due to the assumption that gene expression follows a multivariate normal distribution. Recent studies have revealed that this data in fact follows a super-Gaussian distribution.[39] An alternative called Independent Component Analysis (ICA)[21] has also been proposed to analyze microarray and metabolomics data.[31] As reported in the literature, independent components obtained by ICA are usually better at separating different biological groups than principal components of PCA.[14] However, ICA possesses some limitations related to instability, choice of number of desired component, and handling of high dimensionality. Other examples of methods belonging to this category are Kernel Principle Component Analysis (KPCA),[13] Locality Preserving Projection (LPP),[20] Neighborhood Preserving Embedding (NPE),[19] and Isometric Projection (IsoP).[5]

In parallel to the aforementioned projection-based models, there are attempts to develop a clustering-based counterpart to data transformation. Using of the result of cluster analysis in addition to the native attributes can improve the accuracy of intrusion detection problem.[35] Specific to this study, a fuzzy clustering algorithm is applied to generate memberships of each sample to k clusters (where k is a pre-defined number of clusters). These form additional k variables or dimensions for the following classification process. Similar works have exploited cluster labels and conventional supervised algorithms for the problems of face recognition[33] and image annotation.[41] Presently, an ensemble-based matrix is considered as one variation of transformed data, in which association between samples with respect to cluster memberships are encoded. In particular to the prediction of student dropout,[22,24] the implementation of this approach is examined using the link-based cluster ensemble (LCE) method.[23,26] Based on the comparison with several dimensionality reduction techniques, this has proven more accurate on a number of conventional classifiers. Following this unique idea, the review provides a comparative study of using different graph-based cluster ensemble methods to prepare transformed gene expression data for sample classification. This allows those algorithms invented in the field of ensemble study to be re-used for another important data mining task. Moreover, it illustrates the landscape of predictive performance that is useful for assessing analysis alternatives.

The rest of this paper is organized as follows. Section 3.2 presents the cluster ensemble approach to transforming gene expression data, prior the classification procedure. This includes a review on graph-based cluster ensemble methods that will be explored in the empirical study. Following that, the comparative study on a number of published gene expression datasets and classical classification models is provided in Section 3.3. Discussion regarding the experimental findings and conclusion are shown in Section 3.4, with perspective of future work.

3.2 Cluster Ensemble Approach to Data Transformation

Cluster ensemble or sometimes called ensemble clustering has been at the center of attention for the analysis of microarray data analysis.[25,29] In spite of computational efficiency, this is shown to be an attractive, more accurate choice to many conventional clustering techniques like k-means.[16,43] Based on the previous publication,[25] the problem regarding microarray data can be defined by the followings. Firstly, let $X = \{x_1, \ldots, x_N\}$ be a set of N samples, such that $x_i \in X$ is presented by a vector of d genes or $x_i = (x_{i,1}, \ldots, x_{i,d})$. Moreover, let $\Pi = \{\pi_1, \ldots, \pi_M\}$ be a cluster ensemble with M members or base clusterings. And for each member $\pi_g \in \Pi$, samples are allocated to k_g clusters, i.e., $\pi_g = \{C_1^g, C_2^g, \ldots, C_{k_g}^g\}$ and $\bigcup_{t=1}^{k_g} C_t^g = X$. Provided these, the problem of cluster ensemble is to seek a new clustering that summarizes the information encoded in the ensemble Π. In other words, it is to acquire $\pi^* = C_1^*, \ldots, C_K^*$, where K denotes the number of clusters in the final clustering result. Having set the scene for the common terminology, details of the data transformation framework and graph-based cluster ensemble matrices are included in the next sections.

3.2.1 *Ensemble-Driven Data Transformation Framework*

This framework simply consists of two stages, one for the acquisition of cluster ensemble, and the other for creating cluster information matrices with graph-based cluster ensemble methods.

3.2.1.1 *Ensemble Generation*

As for the current investigation, the following two types of ensembles are examined. According to the original work of LCE,[25] a partitioning algorithm like k-means is used to generate base clusterings, each of which is initialized with a random set of cluster centers or prototypes.

- *Fixed-k*: Each base clustering $\pi_g \in \Pi$, is created using the gene expression dataset $X \in \mathbb{R}^{N \times d}$ with N smaples and d genes. The number of clusters set for each base clustering is fixed to $k = \lceil \sqrt{N} \rceil$.

Note that, to obtain a meaningful data partition, k becomes 50 if $\lceil\sqrt{N}\rceil > 50$.

- *Random-k*: Each base clustering π_g is generated using the gene expression dataset $X \in \mathbb{R}^{N\times d}$ with N smaples and d genes. The number of clusters is randomly selected between $\{2, \ldots, \lceil\sqrt{N}\rceil\}$. Note that both 'Fixed-k' and 'Random-k' generation strategies have become common alternatives for the generation process.[23]

3.2.1.2 *Creating Graph-Based Cluster Ensemble Matrices*

After obtaining the desired ensemble Π of size M, its members are summarized into an information matrix $\Theta_\alpha \in [0,1]^{N\times P}$. Note that P is the total number of clusters in Π (i.e., $P = k_1 + \ldots + k_g$) and α denotes the graph-based method used to create the matrix (i.e., $\alpha \in \{BA, WCT, WTQ, WD\}$). Details of these methods are exhibited below.

- Binary Association (BA) Matrix[15]: each entry in this matrix $\Theta_{BA}(x_i, cl) \in \{0,1\}$ represents a crisp association degree that one sample $x_i \in X$ has with a cluster $cl \in \{\pi_1 \cup \ldots \cup \pi_M\}$.

$$\Theta_{BA}(x_i, cl) = \begin{cases} 1 & \text{if } x_i \in cl \\ 0 & \text{otherwise} \end{cases}, \qquad (1)$$

- Weighted Connected-Triple (WCT) Matrix: from a BA matrix of the target ensemble Π, the entries representing 'nil' associations (i.e., '0') can be approximated from known ones (i.e., '1'), whose association degrees are preserved within the resulting Θ_{WCT} matrix. In other words, $\forall x_i \in X, cl \in \{\pi_1 \cup \ldots \cup \pi_M\}$, $\Theta_{BA}(x_i, cl) = 1 \rightarrow \Theta_{WCT}(x_i, cl) = 1$. For each clustering $\pi_g, g = 1 \ldots M$ and their corresponding clusters $C_1^g, \ldots, C_{k_g}^g$, the association degree $\Theta_{WCT}(x_i, cl) \in [0,1]$ that sample $x_i \in X$ has with each cluster $cl \in \{C_1^g, \ldots, C_{k_g}^g\}$ is estimated by the following equation.

$$\Theta_{WCT}(x_i, cl) = \begin{cases} 1 & \text{if } cl = C_*^g(x_i) \\ sim(cl, C_*^g(x_i)) & \text{otherwise} \end{cases}, \qquad (2)$$

where $C_*^g(x_i)$ is a cluster label to which sample x_i has been assigned. In addition, $sim(C_x, C_y) \in [0,1]$ denotes the similarity between any two clusters $C_x, C_y \in \pi_g$, which can be discovered using a link-based similarity algorithm.

$$sim(C_x, C_y) = \frac{WCT_{xy}}{WCT_{max}} \times DC, \tag{3}$$

here the WCT algorithm[24] is exploited to calculate WCT_{xy} and WCT_{max}, which are the WCT measure between two target clusters (C_x and C_y) and the maximum WCT measure of the entire ensemble, respectively. Note also that $DC \in [0,1]$ is the decay factor that reflects the confidence level of the underlying link analysis.

- Weighted Triple-Quality (WTQ) Matrix: similar to the previous matrix, $\Theta_{WTQ}(x_i, cl) \in [0,1]$ can be estimated by the following.

$$\Theta_{WTQ}(x_i, cl) = \begin{cases} 1 & \text{if } cl = C_*^g(x_i) \\ sim(cl, C_*^g(x_i)) & \text{otherwise} \end{cases}, \tag{4}$$

given that

$$sim(C_x, C_y) = \frac{WTQ_{xy}}{WTQ_{max}} \times DC, \tag{5}$$

here the WTQ algorithm[24] is exploited to calculate WTQ_{xy} and WTQ_{max}, which are the WTQ measure between two target clusters (C_x and C_y) and the maximum WTQ measure of the entire ensemble, respectively.

- Weighted Distance (WD) Matrix: this is developed as the by-product of a new soft-subspace clustering model.[11] An entry $\Theta_{WD}(x_i, cl)$ is estimated from the distance between sample $x_i \in X$ and center of the cluster $cl \in \Pi$. For each base clustering $\pi_g \in \Pi$, $\Theta_{WD}(x_i, cl), \forall cl \in \pi_g$ can be defined as follows.

$$\Theta_{WD}(x_i, cl) = \frac{D_i - d(x_i, \overline{cl}) + 1}{k_g D_i + k_g - \sum_{\forall cl' \in \pi_g} d(x_i, \overline{cl'})}, \tag{6}$$

where $d(x_i, \overline{cl})$ is the distance between sample x_i and $\overline{cl}$, that is center (or centroid) of the cluster cl. In addition, D_i can be specified by the next equation.

$$D_i = \max_{\forall cl \in \pi_g} d(x_i, \overline{cl}). \tag{7}$$

According to Ref. 11, the distance $d(x_i, \overline{cl})$ can be defined as follows.

$$d(x_i, \overline{cl}) = \sqrt{\sum_{s=1}^{d} w_s^{cl}(x_s^i - \overline{cl}_s)^2} \tag{8}$$

provided that $w_s^{cl} \in [0,1]$ is the weight of the sth gene that is specific to the cluster $cl \in \pi_g$, x_s^i denotes value of the sth gene of data sample x_i, and $\overline{cl}_s$ denotes the sth gene value of the cluster center $\overline{cl}$. For any $cl \in \pi_g$,

$$\sum_{s=1}^{d} w_s^{cl} = 1. \tag{9}$$

The set of cluster-specific weights is obtained from a soft subspace clustering such as LAC (Locally Adaptive Clustering[12]). This clustering technique extends the conventional k-means by iteratively revising cluster-specific attribute weights that allow more compact clusters to be achieved.

3.2.2 *Using Ensemble Matrices for Classification*

After generating the transformed matrix $\Theta_\alpha = \{x_1, \ldots, x_N\}$, where $x_i = \{x_{i,1}, \ldots, x_{i,P}\}, i = 1 \ldots N$; it can be exploited to develop a classifier, provided that the ground truth or classes of N samples are known. In other words, the matrix Θ_α can be regarded as the training set for a classification algorithm such as Naive Bayes and C4.5 (Decision Tree). After this supervised-learning phase, the resulting classifier can be used to classify an unseen sample x_u to one of the pre-defined classes. Similar to those in the training set, this new sample is initially presented with the original d genes, i.e., $x_u = \{x_{u,1}, \ldots, x_{u,d}\}$. Since the dimension of this sample vector does not match that of Θ_α, it is mandatory to transform x_u to $x'_u = \{x'_{u,1}, \ldots, x'_{u,P}\}$, which complies to Θ_α. To do this, the knowledge of ensemble Π, WCT-based and WTQ-based similarity measures, as well as the set of cluster-specific weights will be re-used here. The preparation is subjected to the type of ensemble metrix used to create the classifier.

- Case 1: Θ_{BA} is used to create the classifier. For each clustering $\pi_g \in \Pi$, the distances between x_u and centroids of all the clusters belonging to π_g justify the crisp memberships or associations that x_u has with these examined clusters. By repeating this for all clusterings in Π, a Θ_{BA} alike representation of x_u or x'_u is obtained.

- Case 2: Θ_{WCT} is used to create the classifier. This makes use of the resulting x'_u obtained from the first case. In particular, variables (corresponding to clusters) in the vector x'_u with '0' values are modified using the WCT similarity amongst clusters that have been found by the generation of Θ_{WCT}.
- Case 3: Θ_{WTQ} is used to create the classifier. This also makes use of the resulting x'_u obtained from the first case. In particular, variables (corresponding to clusters) in the vector x'_u with '0' values are modified using the WTQ similarity amongst clusters that have been found by the generation of Θ_{WTQ}.
- Case 4: Θ_{WD} is used to create the classifier. For each clustering $\pi_g \in \Pi$, the distances between x_u and centroids of all the clusters belonging to π_g justify the memberships or associations that x_u has with these examined clusters. Reuse the set of cluster-specific weights disclosed during the generation of Θ_{WD}. See details from Equations 6-8. By repeating this for all clusterings in Π, a Θ_{WD} alike representation of x_u or x'_u is acquired.

3.3 Comparative Study of Microarray Data Classification

Having explored the data transformation framework and different ensemble matrices that can be used for classification, it is interesting to observe and compare their performance on different gene expression datasets and experimental settings. The findings from this comparative study can provide insightful information regarding the selection of a graph-based summarization method that is appropriate for a given problem.

3.3.1 *Experimental Design*

To obtain a rigorous comprehension towards the effectiveness of different matrices, this section presents the methodology that is systematically designed and employed for the performance evaluation. At first, this is based on real microarray data obtained from eight published studies. These datasets are summarized in Table 1. The experiments are conducted over filtered datasets as given in the empirical study of de Souto *et al.*[9] where uninformative genes are removed for a better quality of downstreaming analysis.

- Leukemia1-2: These datasets represent details of 72 bone marrow samples that were obtained from acute leukemia patients at the time of diagnosis.[17] They identically contain expression levels of 7,129 genes in Affymetrix high density oligonucleotide chips. In the Leukemia1 dataset, samples are categorized into 47 and 25 cases of acute lymphoblastic leukemia (ALL) and acute myeloid leukemia (AML), respectively. The Leukemia2 dataset presents a refined classification of samples, with the class distribution being 38/9/25 for B-ALL/T-ALL/AML.
- Leukemia3: Samples in this dataset were originally obtained from the peripheral blood or bone marrow of affected individuals at diagnosis or relapse.[2] Each of 72 samples is represented with expression levels of 12,582 genes in Affymetrix chips. In particular, three sample classes are established: 20 cases of lymphoblastic leukemia with MLL translocations (MLL), 24 and 28 conventional acute lymphoblastic (ALL) and acute myelogenous leukemias (AML), respectively.
- Breast-Colon Tumors (BCT): The dataset is based on the original study of Chowdary *et al.* (2006)[7] that compared pairs of snap-frozen and RNAlater preservative-suspended tissue from lymph node-negative breast (B) and Dukes B colon tumors (C). It contains 104 tissue samples each of which is represented by expression levels of 22,283 genes. These samples are grouped into two classes of 62 B and 42 C.
- Brain Tumor: A collection of 50 gliomas were exploited in the investigation of Nutt *et al.* (2003),[36] which examined the methodology to classify high-grade gliomas in a manner more explicit, and consistent than standard pathology. Expression data of 12,625 genes are given for each of the samples that have been categorized into: 14 classic glioblastomas (CG), 14 non-classic glioblastomas (NG), 7 classic anaplastic oligodendrogliomas (CO) and 15 non-classic anaplastic oligodendrogliomas (NO), respectively.
- CNS: Embryonal tumors of the central nervous system (CNS) represent a heterogeneous group of tumors about which little is known biologically. Yet their diagnosis, based on morphologic appearance alone, is controversial. To overcome such problem, a classification systems based on gene expression data was developed in Pomeroy

et al. (2002).[38] In particular, a collection of experimented samples includes 10 cases of medulloblastomas (MD), 8 primitive neuroectodermal tumors (PNET), 10 atypical teratoid/rhabdoid tumors (Rhab), 10 malignant gliomas (Mglio) and 4 normal tissues.
- HCC: Hepatocellular carcinoma (HCC) is the most common liver malignancy and among the five leading causes of cancer death worldwide. In the study of Chen *et al.* (2002),[6] cDNA microarrays were exploited to compare patterns of gene expression in HCC and those in non-tumor liver tissues (LIVER). This dataset contains 180 samples (104 HCC and 76 LIVER), each of which is presented by expression levels of 22,699 genes.
- SRBCT: Small, round blue-cell tumors (SRBCTs) were used with artificial neural networks to develop a method of classifying cancers to specific diagnostic categories.[28] The underlying 83 cancer samples belong to one of neuroblastomas (NB), Burkitt lymphoma (BL), rhabdomyosarcoma (RMS) and the Ewing family of tumors (EWS). In addition, the class distribution is 29 EWS, 11 BL, 18 NB and 25 RMS.

Besides, experimental settings are explained below.

- k-means is used to generate base clusterings in case of Θ_{BA}, Θ_{WCT} and Θ_{WTQ}; while LAC is used for the preparation of Θ_{WD}. The ensemble size $M = 10$ is assessed.
- The value of DC is set to 0.9 for the generation of Θ_{WCT} and Θ_{WTQ} matrices. For each dataset, the generation of ensemble and corresponding matrix is repeated for 20 trials.
- Three conventional techniques C4.5 (Decision Tree), Naive Bayes and KNN (K=1) are used to generate classification models from the original and investigated data matrices. Given a data matrix, 10-fold cross validation is specifically exploited to determine the classification error rate $\in [0, 1]$, where error rate of 0 indicates the most accurate case with no false positive nor false negative.

3.3.2 *Experimental Results*

As for the application of Naive Bayes to different matrices, Table 2 presents dataset-specific results obtained with Fixed-k (FK) and Random-k (RK) generation schemes. Both $\Theta_{WCT}(FK)$ and $\Theta_{WTQ}(FK)$ often provide

Table 1. Description of gene expression datasets: tissue type, microarray chip type, number of samples (N), number of original genes (d^*), number of selected genes (d) after pre-processing, and number of classes (K).

Dataset	Tissue	Chip	Samples (N)	Original genes (d^*)	Selected genes (d)	Classes (K)
Leukemia1[17]	Bone marrow	Affy	72	7,129	1,877	2
Leukemia2[17]	Bone marrow	Affy	72	7,129	1,877	3
Leukemia3[2]	Blood	Affy	72	12,582	2,194	3
BCT[7]	Breast and Colon	Affy	104	22,283	182	2
Brain Tumor[36]	Brain	Affy	50	12,625	1,377	4
CNS[38]	Brain	Affy	42	7,129	1,379	5
HCC[6]	Liver	cDNA	180	22,699	85	2
SRBCT[28]	Multi-tissue	cDNA	83	6,567	1,069	4

more accurate outcome than their RK counterparts and the other matrices. Despite this, $\Theta_{WCT}(RK)$ and $\Theta_{WTQ}(RK)$ have shown exceptional results in a fews cases such as the CNS dataset. In addition, the two variations of Θ_{WD} are effective for the analysis of Leukemia3 and BCT, while Θ_{BA} usually has higher error rates than the rest. Similar trends can also be observed with the applications of C4.5 and KNN classification models to these matrices, with the corresponding statistics being illustrated in Tables 3 and 4 , respectively. Note that the classification performance with Θ_{BA} matrices have improved with these two classifiers, while those of Θ_{WD} become of lower quality.

To further elaborate the empirical findings, Figure 1 shows for the case of Naive Bayes the comparison of error rates as the averages across investigated datasets. For the four methods to generate Θ (WCT, WTQ, BA and WD) and two generation schemes (FK and RK), the best results are equally obatined by WCT(FK) and WTQ(FK), and the two worst still occur with BA matrices. The FK strategy is generally better than the other with WCT, WTQ and WD, while it is the other way round for BA. For the summarization of C4.5 and KNN, Figures 2 and 3 similarly suggest that WCT(FK) and WTQ(FK) are the most accurate, which are slightly better than the BA(FK) alternative. Figure 4 provides the results with KNN classifiers, where the number of neighbors of K varies from 1 to 3. Unlike the previous observation with Naive Bayes, WD matrices are less effective with the models of C4.5 and KNN. Based on these, the original BA matrix can usually represent the knowledge embedded in an ensemble rather well, with a possible improvement by a robust matrix refinement approach, e.g., link-based methods. However, the distance-orient refining technique like

Table 2. Classification errors with Naive Bayes, where FK and RK denotes Fixed-k and Random-k generation strategies. The two lowest error rates on each investigated dataset is highlighted in **boldface**.

Dataset	Θ_{WCT} (FK)	Θ_{WCT} (RK)	Θ_{WTQ} (FK)	Θ_{WTQ} (RK)	Θ_{BA} (FK)	Θ_{BA} (RK)	Θ_{WD} (FK)	Θ_{WD} (RK)
Leukemia1	**0.107**	0.160	**0.142**	0.196	0.374	0.335	0.158	0.160
	(0.051)	(0.064)	(0.078)	(0.093)	(0.031)	(0.074)	(0.031)	(0.046)
Leukemia2	**0.186**	0.300	**0.214**	0.311	0.567	0.412	0.224	0.235
	(0.045)	(0.100)	(0.075)	(0.095)	(0.044)	(0.104)	(0.040)	(0.047)
Leukemia3	**0.094**	0.125	0.119	0.129	0.338	0.322	**0.082**	0.125
	(0.050)	(0.063)	(0.066)	(0.071)	(0.008)	(0.039)	(0.022)	(0.032)
BCT	0.105	0.111	0.105	0.081	0.299	0.099	**0.063**	**0.060**
	(0.030)	(0.044)	(0.030)	(0.047)	(0.083)	(0.028)	(0.053)	(0.034)
Brain Tumor	**0.355**	0.415	**0.364**	0.467	0.709	0.527	0.450	0.446
	(0.079)	(0.061)	(0.087)	(0.069)	(0.018)	(0.053)	(0.048)	(0.063)
CNS	0.483	0.379	0.462	**0.364**	0.411	0.417	0.385	**0.371**
	(0.091)	(0.091)	(0.110)	(0.085)	(0.040)	(0.045)	(0.052)	(0.076)
HCC	0.101	0.104	**0.083**	**0.084**	0.419	0.411	0.125	0.123
	(0.022)	(0.029)	(0.013)	(0.015)	(0.000)	(0.025)	(0.037)	(0.041)
SRBCT	**0.250**	0.320	**0.243**	0.292	0.675	0.667	0.330	0.316
	(0.071)	(0.085)	(0.056)	(0.089)	(0.029)	(0.047)	(0.059)	(0.072)

Table 3. Classification errors with C4.5, where FK and RK denotes Fixed-k and Random-k generation strategies. The two lowest error rates on each investigated dataset is highlighted in **boldface**.

Dataset	Θ_{WCT} (FK)	Θ_{WCT} (RK)	Θ_{WTQ} (FK)	Θ_{WTQ} (RK)	Θ_{BA} (FK)	Θ_{BA} (RK)	Θ_{WD} (FK)	Θ_{WD} (RK)
Leukemia1	0.094	**0.090**	0.099	**0.090**	0.105	0.096	0.110	0.106
	(0.052)	(0.038)	(0.045)	(0.043)	(0.058)	(0.050)	(0.020)	(0.019)
Leukemia2	**0.109**	0.121	0.111	0.119	**0.110**	0.126	0.229	0.229
	(0.046)	(0.041)	(0.045)	(0.044)	(0.043)	(0.051)	(0.026)	(0.030)
Leukemia3	0.072	**0.071**	**0.071**	0.070	0.087	0.083	0.074	0.072
	(0.023)	(0.030)	(0.022)	(0.028)	(0.021)	(0.038)	(0.031)	(0.024)
BCT	**0.021**	**0.021**	0.022	**0.021**	0.039	0.031	0.051	0.053
	(0.004)	(0.003)	(0.007)	(0.003)	(0.022)	(0.004)	(0.008)	(0.008)
Brain Tumor	0.276	0.282	**0.266**	**0.274**	0.279	0.283	0.332	0.334
	(0.040)	(0.043)	(0.035)	(0.054)	(0.040)	(0.046)	(0.031)	(0.048)
CNS	**0.270**	0.310	**0.276**	0.312	0.312	0.312	0.286	0.288
	(0.050)	(0.043)	(0.047)	(0.050)	(0.046)	(0.054)	(0.035)	(0.033)
HCC	**0.076**	**0.077**	0.080	0.080	0.088	0.080	0.124	0.111
	(0.015)	(0.014)	(0.016)	(0.012)	(0.020)	(0.013)	(0.035)	(0.032)
SRBCT	**0.151**	0.259	**0.147**	0.250	**0.151**	0.251	0.271	0.282
	(0.039)	(0.089)	(0.045)	(0.082)	(0.035)	(0.087)	(0.078)	(0.057)

Table 4. Classification errors with KNN, where FK and RK denotes Fixed-k and Random-k generation strategies. The two lowest error rates on each investigated dataset is highlighted in **boldface**.

Dataset	Θ_{WCT} (FK)	Θ_{WCT} (RK)	Θ_{WTQ} (FK)	Θ_{WTQ} (RK)	Θ_{BA} (FK)	Θ_{BA} (RK)	Θ_{WD} (FK)	Θ_{WD} (RK)
Leukemia1	0.094	**0.088**	**0.092**	0.094	0.093	0.097	0.236	0.233
	(0.034)	(0.035)	(0.037)	(0.036)	(0.038)	(0.038)	(0.026)	(0.023)
Leukemia2	0.118	0.130	**0.116**	0.128	**0.115**	0.126	0.439	0.447
	(0.038)	(0.074)	(0.036)	(0.074)	(0.038)	(0.074)	(0.024)	(0.039)
Leukemia3	**0.174**	0.211	**0.173**	0.213	0.176	0.215	0.200	0.201
	(0.067)	(0.050)	(0.067)	(0.051)	(0.064)	(0.048)	(0.008)	(0.009)
BCT	**0.111**	0.163	**0.108**	0.163	0.118	0.162	0.163	0.165
	(0.022)	(0.076)	(0.021)	(0.077)	(0.021)	(0.078)	(0.034)	(0.023)
Brain Tumor	**0.504**	0.569	**0.506**	0.569	0.509	0.580	0.576	0.572
	(0.060)	(0.082)	(0.058)	(0.080)	(0.053)	(0.072)	(0.049)	(0.061)
CNS	**0.274**	0.301	**0.271**	0.298	0.281	0.291	0.488	0.495
	(0.052)	(0.062)	(0.049)	(0.061)	(0.051)	(0.062)	(0.040)	(0.044)
HCC	0.090	0.099	**0.087**	0.100	**0.088**	0.099	0.198	0.196
	(0.020)	(0.043)	(0.022)	(0.045)	(0.019)	(0.044)	(0.044)	(0.035)
SRBCT	**0.225**	0.292	**0.215**	0.287	0.234	0.292	0.353	0.355
	(0.034)	(0.069)	(0.037)	(0.073)	(0.030)	(0.069)	(0.038)	(0.034)

WD appears not to always be reliable as such, at least for the analysis of microarray data.

3.4 Conclusion

This paper has presented the comparative of graph-based ensemble clustering as transformation methods for microarray data classification. The problem and analysis framework was initially defined with several methods being specified for the preparation of desired data. These include the original matrix usually used to summarize a cluster ensemble (i.e., BA) and its refined variations, which are obtained using a link-based or distance-based approach. To justify the use of different matrices for microarray data classification, a number of published datasets are included in the empirical study, where a generalized experimental setting is employed to derive a robust finding.

The results with three different classifiers suggest that BA is a generally good alternative to transform microarray data. In particular, its refined variations using link-based techniques can raise the classification accuracy further. On the other hand, the refinement with a distance-based technique is less effective and pretty much subjected to the data under examination.

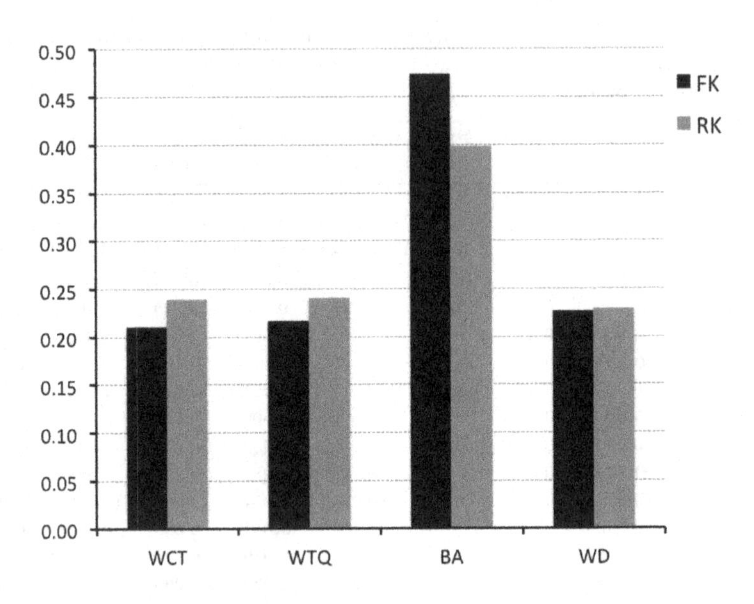

Fig. 1. Comparison of classification errors (y-axis) as averages across all datasets with Naive Bayes, categorized by matrix type and generation strategy.

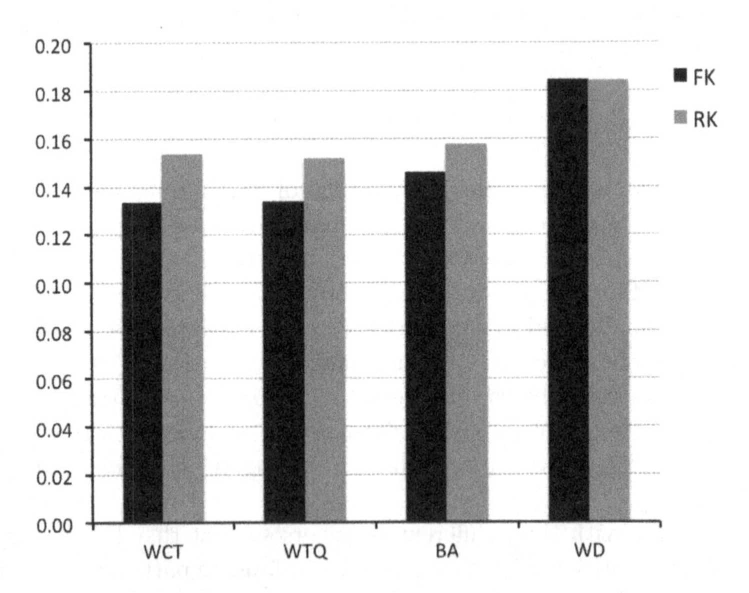

Fig. 2. Comparison of classification errors (y-axis) as averages across all datasets with C4.5, categorized by matrix type and generation strategy.

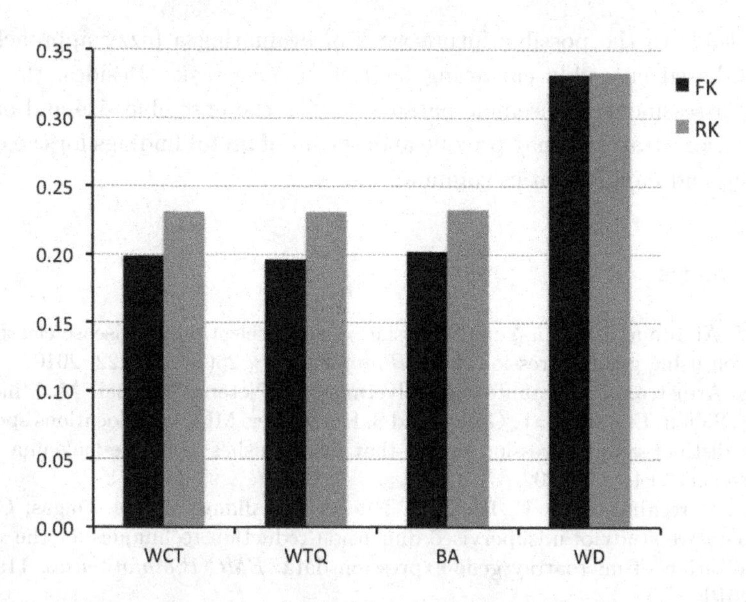

Fig. 3. Comparison of classification errors (y-axis) as averages across all datasets with KNN, categorized by matrix type and generation strategy.

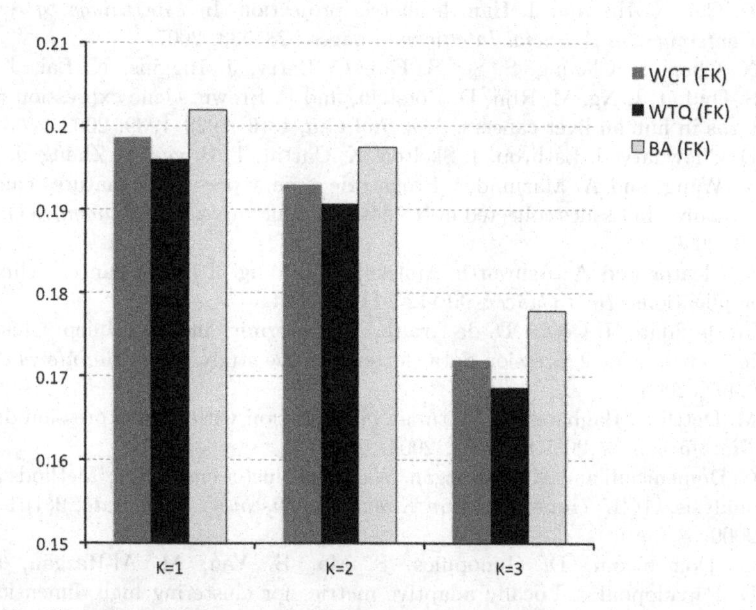

Fig. 4. Comparison of classification errors (y-axis) as averages across all datasets with KNN and different K values from 1 to 3; categorized by three specific matrices of WCT(FK), WTQ(FK) and BA(FK), respectively.

This leads to the possible future work of evaluating a fuzzy approach to graph-based ensemble clustering for this specific task. Besides, the rigorous assessment of ensemble parameters like the ensemble size and other generation strategies may provide another set of useful findings for the data mining and bioinformatics communities.

References

1. Y. Ai-Jun and S. Xin-Yuan. Bayesian variable selection for disease classification using gene expression data. *Bioinformatics*, 26(2):215–222, 2010.
2. S. Armstrong, J. Staunton, L. Silverman, R. Pieters, M. Boer, M. Minden, S. Sallan, E. Lander, T. Golub, and S. Korsmeyer. MLL translocations specify a distinct gene expression profile that distinguishes a unique leukemia. *Nat Genet*, 30:41–47, 2002.
3. C. Bartenhagen, H. U. Klein, C. Ruckert, X. Jiang, and M. Dugas. Comparative study of unsupervised dimension reduction techniques for the visualization of microarray gene expression data. *BMC Bioinformatics*, 11:567, 2010.
4. T. Bo and I. Jonassen. New feature subset selection procedures for classification of expression profiles. *Genome Biology*, 3:1–17, 2002.
5. D. Cai, X. He, and J. Han. Isometric projection. In *Proceedings og AAAI Conference on Artificial Intelligence*, pages 528–533, 2007.
6. X. Chen, S. Cheung, S. So, S. Fan, C. Barry, J. Higgins, K. Lai, J. Ji, S. Dudoit, I. Ng, M. Rijn, D. Botstein, and P. Brown. Gene expression patterns in human liver cancers. *Mol Biol Cell*, 13(6):1929–1939, 2002.
7. D. Chowdary, J. Lathrop, J. Skelton, K. Curtin, T. Briggs, Y. Zhang, J. Yu, Y. Wang, and A. Mazumder. Prognostic gene expression signatures can be measured in tissues collected in RNAlater preservative. *J Mol Diagn*, 8(1):31–39, 2006.
8. S. Cleator and A. Ashworth. Molecular profiling of breast cancer: clinical implications. *Br J Cancer*, 90:1120–1124, 2004.
9. M. de Souto, I. Costa, D. de Araujo, T. Ludermir, and A. Schliep. Clustering cancer gene expression data: a comparative study. *BMC Bioinformatics*, 9:497, 2008.
10. M. Dettling. Bagboosting for tumor classification with gene expression data. *Bioinformatics*, 20:3583–3593, 2004.
11. C. Domeniconi and M. Al-Razgan. Weighted cluster ensembles: Methods and analysis. *ACM Transactions on Knowledge Discovery from Data*, 2(4):1–40, 2009.
12. C. Domeniconi, D. Gunopulos, S. Ma, B. Yan, M. Al-Razgan, and D. Papadopoulos. Locally adaptive metrics for clustering high dimensional data. *Data Mining and Knowledge Discovery*, 14(1):63–97, 2007.
13. R. O. Duda, P. E. Hart, and D. G. Stork. *Pattern Classification*. Wiley-Interscience, 2012.

14. J. Engreitz, B. Jr. Daigle, J. Marshall, and R. Altman. Independent component analysis: Mining microarray data for fundamental human gene expression modules. *Journal of Biomedical Informatics*, 43:932–944, 2010.
15. X. Z. Fern and C. E. Brodley. Solving cluster ensemble problems by bipartite graph partitioning. In *Proceedings of International Conference on Machine Learning*, pages 36–43, 2004.
16. A. L. N. Fred and A. K. Jain. Combining multiple clusterings using evidence accumulation. *IEEE Trans. on Pattern Analysis and Machine Intelligence*, 27(6):835–850, 2005.
17. T. Golub, D. Slonim, P. Tamayo, C. Huard, M. Gaasenbeek, J. Mesirov, H. Coller, M. Loh, J. Downing, M. Caligiuri, C. Bloomfield, and E. Lander. Molecular classification of cancer: class discovery and class prediction by gene expression monitoring. *Science*, 286:531–537, 1999.
18. M. Gupta and J. G. Ibrahim. Variable selection in regression mixture modeling for the discovery of gene regulatory networks. *Journal of American Statistics Association*, 102:867–880, 2007.
19. X. He, D. Cai, S. Yan, and H. J. Zhang. Neighborhood preserving embedding. In *Proceedings of International Conference on Computer Vision*, pages 1208–1213, 2005.
20. X. He, S. Yan, Y. Hu, P. Niyogi, and H. J. Zhang. Face recognition using laplacianfaces. *IEEE Transactions on Pattern Analysis and Machine Intelligence*, 27(3):328–340, 2005.
21. A. Hyvarinen and E. Oja. Independent component analysis: Algorithms and applications. *Neural Networks*, 13(4-5):411–430, 2000.
22. N. Iam-on and T. Boongoen. Revisiting link-based cluster ensembles for microarray data classification. In *Proceedings of IEEE International Conference on Systems, Man and cybernetics*, pages 4543–4548, 2013.
23. N. Iam-on and T. Boongoen. Comparative study of matrix refinement approaches for ensemble clustering. *Machine Learning*, 98(1-2):269–300, 2015.
24. N. Iam-on and T. Boongoen. Diversity-driven generation of link-based cluster ensemble and application to data classification. *Expert Systems with Applications*, 42(21):8259–8273, 2015.
25. N. Iam-on, T. Boongoen, and S. Garrett. LCE: A link-based cluster ensemble method for improved gene expression data analysis. *Bioinformatics*, 26(12):1513–1519, 2010.
26. N. Iam-on, T. Boongoen, and N. Kongkotchawan. A new link-based method to ensemble clustering and cancer microarray data analysis. *International Journal of Collaborative Intelligence*, 1(1):45–67, 2014.
27. I. Joliffe. *Principal Component Analysis*. Springer: New York, 1986.
28. J. Khan, J. Wei, M. Ringner, L. Saal, M. Ladanyi, F. Westermann, F. Berthold, M. Schwab, C. Antonescu, C. Peterson, and P. Meltzer. Classification and diagnostic prediction of cancers using gene expression profiling and artificial neural networks. *Nat. Med.*, 7(6):673–679, 2001.
29. E. Kim, S. Kim, D. Ashlock, and D. Nam. MULTI-K: accurate classification

of microarray subtypes using ensemble k-means clustering. *BMC Bioinformatics*, 10:260, 2009.

30. K. E. Lee, N. Sha, E. R. Dougherty, M. Vannucci, and B. K. Mallick. Gene selection: a Bayesian variable selection approach. *Bioinformatics*, 19:90–97, 2003.
31. S. Lee and S. Batzoglou. Application of independent component analysis to microarrays. *Genome Biology*, 4(11):R76, 2003.
32. H. Mamitsuka. Selecting features in microarray classification using roc curves. *Pattern Recognition*, 39(12):2393–2404, 2006.
33. G. Nasierding, G. Tsoumakas, and A. Z. Kouzani. Clustering based multi-label classification for image annotation and retrieval. In *Proceedings of IEEE International Conference on System, Man and Cybernetics*, pages 4514–4519, 2009.
34. D. V. Nguyen and D. M. Rocke. Tumor classification by partial least squares using microarray gene expression data. *Bioinformatics*, 18:39–50, 2002.
35. H. H. Nguyen, N. Harbi, and J. Darmont. An efficient fuzzy clustering-based approach for intrusion detection. In *Proceedings of IEEE International Conference on Data Mining*, pages 607–612, 2011.
36. C. Nutt, D. Mani, R. Betensky, P. Tamayo, J. Cairncross, C. Ladd, U. Pohl, C. Hartmann, M. McLaughlin, T. Batchelor, P. Black, A. Deimling, S. Pomeroy, T. Golub, and D. Louis. Gene expression based classification of malignant gliomas correlates better with survival than histological classification. *Cancer Res*, 63(7):1602–1607, 2003.
37. W. Pan. A comparative review of statistical methods for discovering differentially expressed genes in replicated microarray experiments. *Bioinformatics*, 18:546–554, 2002.
38. S. Pomeroy, P. Tamayo, M. Gaasenbeek, L. Sturla, M. Angelo, M. McLaughlin, J. Kim, L. Goumnerova, P. Black, C. Lau, J. Allen, D. Zagzag, J. Olson, T. Curran, C. Wetmore, J. Biegel, T. Poggio, S. Mukherjee, R. Rifkin, A. Califano, G. Stolovitzky, D. Louis, J. Mesirov, E. Lander, and T. Golub. Prediction of central nervous system embryonal tumour outcome based on gene expression. *Nature*, 415(6870):436–442, 2002.
39. E. Purdom and S. Holmes. Error distribution for gene expression data. *Statistical applications in genetics and molecular biology*, 4:16, 2005.
40. S. Ramaswamy, K. Ross, E. Lander, and T. Golub. A molecular signature of metastasis in primary solid tumors. *Nat Genet*, 33:49–54, 2003.
41. K. Sang-Woon. A pre-clustering technique for optimizing subclass discriminant analysis. *Pattern Recognition Letters*, 31(6):462–468, 2010.
42. R. Spang. Diagnostic signatures from microarrays: a bioinformatics concept for personalized medicine. *BIOSILICO*, 1:264–268, 2003.
43. A. P. Topchy, A. K. Jain, and W. F. Punch. Clustering ensembles: Models of consensus and weak partitions. *IEEE Transaction on Pattern Analysis and Machine Intelligence*, 27(12):1866–1881, 2005.
44. V. Tusher, R. Tibshirani, and G. Chu. Significance analysis of microarrays applied to the ionizing radiation response. *Proc Natl Acad Sci USA*, 98(9):5116–5121, 2001.

45. A. Wallqvist, A. Rabow, R. Shoemaker, E. Sausville, and D. Covell. Establishing connections between microarray expression data and chemotherapeutic cancer pharmacology. *Mol Cancer Ther*, 1:311–320, 2002.
46. A. Wang, N. An, G. Chen, L. Li, and G. Alterovitz. Tumor classification by combining pnn classifier ensemble with neighborhood rough set based gene reduction. *Computers in Biology and Medicine*, 62:14–24, 2015.
47. S. L. Wang, X. Li, S. Zhang, J. Gui, and D. S. Huang. Tumor classification by combining pnn classifier ensemble with neighborhood rough set based gene reduction. *Computers in Biology and Medicine*, 40(2):179–189, 2010.

Chapter 4

Semantic analytics of biomedical data[a]

Charles C.N. Wang[†,*], Phillip C.-Y. Sheu[†,‡] and Jeffrey J. P. Tsai[†]
[†]*Department of Biomedical Informatics, Asia University, 500, Lioufeng Rd., Wufeng, Taichung 41354, Taiwan*
[‡]*Department of Electrical Engineering and Computer Science, University of California, Irvine, 5200 Engineering Hall, Irvine, CA 92697, USA*
[*]*chaoneng.wang@gmail.com*

Biomedical intelligence (BMI) has been studied in solos, lacking a systematic methodology. Bioinformatics has been conceptualizing biological process in terms of genomics and applying computer science (derived from disciplines such as applied modeling, data mining, machine learning and statistics) to extract knowledge from biological data. Medical Informatics, on the other hand, has been developing health care applications based on clinical observations and applying computer science to extract knowledge and information to facilitate problem solving and decision marking In this chapter, we describe how semantic computing can enhance biological and medical intelligence. Specifically, we show how structured natural language (SNL) can express many problems in BMI with a finite number of sentence patterns, and show how biological analysis tools, OLAP, data mining and statistical analysis may be linked to solve problems related to biomedical data.

Keywords: Biomedical intelligence, Semantic Computing, Structured Natural Language, Bioinformatics, Medical Informatics.

---

[a]A part of this chapter is revised from Charles C. N. Wang, Phillip C.-Y. Sheu, and Jeffrey J. P. Tsai, Towards Semantic Biomedical Problem Solving, Int. J. Semantic Computing, Vol 09, pp 415 (2015).

4.1 Introduction

Leveraging the technological advancements in molecular biology and genetics in the 1980s, the Human Genome Project (HGP) was initiated with the goal of enabling progress and benefits in biomedicine. The HGP published that new challenges in functional genomics have followed hard on its heels, opening up a wide variety of medical applications (Lander *et al.*, 2001). In recent years the need for new genomic approaches in medicine have emerged, such as finding genome related risk factors for disease, creating updated cancer cell classification and integrating genetic and medical data in clinical practice. Bioinformatics and Medical informatics are widely expected to have important roles in supporting these types of efforts, but whether they will do so together or apart has been debated among researchers in both disciplines. Bioinformatics is conceptualizing biological process in terms of genomics and applying computer science (derived from disciplines such as applied modeling, data mining, machine learning and statistics) to extract knowledge from biological data. And The term Medical Informatics was introduced as a MeSH term in 1987 (Lowe and Barnett, 1994). Medical Informatics expertise in developing health care applications and the strength of bioinformatics in biological discovery science complement each other well. Medical informatics is conceptualizing clinical observations and applying computer science to extract knowledge, information, problem solving and decision marking from medical data (Hristovski, Dinevski, Kastrin, and Rindflesch, 2015a).

On another front, Semantic Computing has been drawing more and more attention in academia and industries. It brings together various analytics techniques to connect the (often vaguely formulated) intentions of humans with computational content that includes, but is not limited to, structured and semi-structured data, multimedia data, text, etc (Sheu *et al.*, 2011). Dimitar (Hristovski, Dinevski, Kastrin, and Rindflesch, 2015b) proposes a semantic methodology and describes a tool called SemBT for biomedical question answering. The system is able to provide answers to a wide array of questions, from clinical medicine through pharmacogenomics to microarray results interpretation. Zhang (Zhang *et al.*, 2014) presents a semantic methodology for detecting drug-drug interactions in clinical data that exploits semantic predications. The results suggest that the use of structured knowledge in the form of relationships from biomedical literatures can support the discovery of potential drug-drug interactions occurring in patient clinical data.

In this chapter, we describe how Semantic Computing can enhance BMI. Specifically, we show how Structured Natural Language (SNL) can express many problems in BMI with a finite number of sentence patterns, and show how OLAP and data mining tools may be linked to solve problems related to biomedical data. In addition, we show how a language-based analysis of a problem domain may help to discover new problems.

4.2 Related Work

Bioinformatics involves the development and application of novel informatics techniques in biological (especially genomic) sciences. It is a young and successful discipline, which already has its own professional societies, meetings, and scientific journals focused on a clear research agenda, having contributed critical to the successes of the human and other genome projects. In contrast, medical informatics is a more established field that has pioneered the development and introduction of informatics methods in clinical medicine and biomedical research but has recently found itself increasingly challenged by the emergence of bioinformatics (Maojo and Kulikowski, 2003).

On the other hand, Business Intelligence (BI) includes a set of techniques and tools for the transformation of raw data into meaningful and useful information for business analysis (Rud, 2012). Business analytics is the practice of iterative, methodical exploration of raw data with an emphasis on statistical analysis. It's used by companies committed to data-driven decision making to gain insights of data and used them to automate and optimize business processes. Data-driven companies treat their data as a corporate asset and leverage it for competitive advantages. Successful business analytics depends on data quality, skilled analysts who understand the technologies and the business, and an organizational commitment to data-driven decision making. BI tools are designed to retrieve, analyze, transform and report data. The general categories of business intelligence tools include online analytical processing (OLAP) and data mining. OLAP is a powerful technique for analyzing business problems in terms of a multidimensional model. Traditional OLAP models (such as ROLAP and MOLAP) are not sufficient to answer queries for complex applications (Ikeda *et al.*, 2011). Data mining is a process of knowledge discovery involving finding hidden patterns and associations, constructing analytical models, performing classifications, and predictions (Data Mining: Concepts and Techniques, 2011) .

We may define Biomedical Intelligence (BMI) to include bioinformatics tools that can be used to transform biological data into meaningful and useful information, such as sequence alignment, gene finding, genome assembly, analysis of differential expressions, protein structure alignment and prediction of gene regulatory networks, and include techniques and tools for the transformation of medical data into meaningful and useful information, such as identification of comorbidity, assessment of risks and prediction of diseases. BMI technologies are expected to be capable of handling large amounts of unstructured as well as structured data. A goal of medical intelligence is to allow for easy interpretation and understanding of large medical datasets.

4.3 Structured Natural Language

We shall describe a method to description BMI problems in *SemanticObject*s which is an object-relational framework that, by providing an object relational layer on top of the relations in a relational database, accounts for a seamless integration of information presented at various levels together with associated tools (algorithms) with the uniform concept of object (Sheu and Kitazawa, 2007). This allows for many abstract concepts or domain dependent methods that cannot be expressed in the relational model to be included as a part of a query.

Queries and commands in *SemanticObjects* are structured along the lines of Structured Natural Language (SNL):

SNL: Verb Nouns conditions

in which objects (nouns) are identified, described (adjectives, predicates) and acted upon (verbs). Naïve users can compose queries based on simple multiple hierarchical choices without knowing any low-level concepts such as "join" and "selection", e.g., Correlate patients who are 1 to 5 years old with disease ADHD. The query consists of one verb ("correlate"), one noun ("patients") and two conditions ("patients are 1 to 5 years old" and "with disease ADHD").

4.4 Problem Spaces for Biological and Medical Intelligence

In this section, we show how Structured Natural Language (SNL) can express many problems in BMI with a finite number of sentence patterns, and show how biological analysis tools, systems biology, OLAP, data mining tools and statistical analysis tools may be linked to solve problems with a uniform interface.

4.4.1 *Problem Space for Biological Intelligence*

Bioinformatics tools have been developed for biological analytics applications such as sequence analysis, gene expression analysis, protein structural predication, biological network and systems biology(Tarczy-Hornoch and Minie, 2005). In our study, we assume the analyst interacts with the biological intelligence system with respect to:

Sequence analysis
Gene expression analysis
Protein structure prediction
Biological network and Computational systems biology

In the following sections we shall illustrate the use of SNL for describing some typical biological applications.

4.4.1.1 *Sequence Analysis*

Sequence analysis of gene and proteins represents a fundamental class of applications that are routinely preformed. These analyses depend solely on the underlying nucleic acid sequences for genes and the amino acid sequences for proteins. Stephen *et al.* (Altschul *et al.*, 1990) propose an approach to rapid sequence comparison in 1990, call the basic local alignment search tool (BLAST), that directly approximates alignments that optimize a measure of local similarity, the maximal segment pair score. The basic algorithm is simple and robust; it can be implemented in a number of ways and be applied in a variety of contexts including DNA, RNA and protein sequence database search, motif search, gene identification search, and in the analysis of multiple regions of similarity in long sequences. BLAST can be used for several purposes.

1) Identifying species: BLAST can correctly identify a species or find homologous species.
2) Locating domains: BLAST can help locate known domains within the sequence of interest.
3) DNA mapping: When working with a known species, and trying to sequence a gene at an unknown location, BLAST can compare the chromosomal position of the sequence of interest to other relevant sequences in the species database.

Presented below are several case studies chosen to demonstrate how sequencing, protein structural analysis and alignment problems may be described in SNL.

BLAST Problems

Perhaps one of the most common tasks in biological research today is that of identifying genes and proteins related or similar to a particular sequence. The task is often performed with BLAST. A representative problem is presented below, where variable parameters are preceded with the dollar sign '$':

Nucleotide BLAST

- Search $nucleotide-database using $nucleotide

Protein BLAST
- Search $protein-database using $protein

BLASTX
- Search $protein-database using $translated-nucleotide

TBLASTN
- Search $translated-nucleotide-database using $protein

TBLASTX
- Search $translated-nucleotide-database using $translated-nucleotid

A specialized BLAST problem is sequence alignment. The objective is to identify which regions are conserved and which are different. This problem becomes complicated by the fact that there can be intervening sequences of varying lengths that play little or no functional/structural role. The SNL sentence describing the sequence alignment problem is:

Nucleotide BLAST
- Align $nucleotide and $nucleotide

The problem may be solved using one or more of the following tools:

- NCBI BLAST: http://blast.ncbi.nlm.nih.gov/Blast.cgi
- EBI BLAST: http://www.ebi.ac.uk/Tools/msa/

4.4.1.2 *Gene Expression Analysis*

Gene expressions have been widely used in the synthesis of functional gene products. Microarrays can simultaneously measure the expression level of thousands of genes within a particular mRNA sample. Microarrays are routinely used to study gene expressions and gene regulatory networks. They are also increasingly being used to identify biomarkers and validate drug targets, as well as to study the gene and potential toxicological effects of compounds in a model (Fryer *et al.*, 2002).

Microarray experiments are routinely used to study gene expressions and gene regulatory networks. They are also increasingly being used to identify biomarkers and to validate drug targets, as well as to study the gene and potential toxicological effects of compounds in a model (MACGREGoR and SqUIRE, 2002). The result of a microarray analysis is usually presented as a list of genes whose expressions are considered to change (and they are known as differentially expression genes). The identification of differential gene expressions, clustering and classification are the task of an in depth microarray analysis.

For a microarray analysis, clustering is often the first step to be performed; it employs an unsupervised approach to classify the genes into groups with similar patterns. Classification is then performed with a supervised learning method; it is also known as class prediction or discriminant analysis. Generally, classification is a process of learning-from-examples (Mutch *et al.*, 2001). The SNL sentences describing some representative problems in microarray analysis are presented below.

Normalization

Current microarray studies often only contain a small number of microarray experiments data, resulting in limited robustness and reliability for statistical analysis. In microarray experiments, there are many sources of systematic variations. Normalization is the process of removing some sources of variations which affect the measured gene expression levels. It play an important role in the earlier stage of microarray analysis. Existing normalization methods include: (1) an approach based on linked gene and gene clustering (XPN), (2) an empirical Bayes method, (3) MRANK which is a median rank score based method, (4) an outlier removing discretization technique (NorDi), and (5) a quintile discretization procedure.

Normalization of microarray experiments

- Normalize $microarray-experiments

Identification of Differentially Expressed Genes

Identifying differential expression genes is a common starting point for the biological interpretation of microarray data. Microarray data allows the simultaneous monitoring of thousands of gene expressions.

Several differential gene expression analysis methods have been used to determine either the expression or relative expression of a gene from normalized microarray data, including (1) empirical Bayes t-statistic (Kim *et al.*, 2006), (2) significance analysis (Tusher *et al.*, 2001), (3) correlation based combinatorial feature selection (Hall, 2000), and (4) partial least Square based filter (Boulesteix and Strimmer, 2007).

Identifying differential expressed genes - Find $differential-genes based on $microarray-experiments

Data Mining for Microarray

In recent years, microarray data mining methods such as clustering, classification and association analysis heavily rely on statistical analysis of gene expression data. Microarray data mining may be the most popular method currently used. It is used for finding co-regulated, functionally related groups and finding the rules that allow assigning new samples to one of the classes (Svrakic *et al.*, 2003). For microarray analysis, clustering analysis has been playing a growing role in the study of co-expressed genes for gene discovery. Clustering is often the first step to be performed; it employs an unsupervised approach to classify the genes into groups with similar patterns. Classification is then performed with a supervised learning method; it is also known as class prediction or discriminant analysis. Data mining methods have also been actually proposed to address various issues specific to gene discovery problems such as consistent co-expression of genes over multiple microarray datasets (Abu-Jamous *et al.*, 2013) (Natarajan, 2013). The common clustering methods include hierarchical clustering and k-means clustering, and the common classification methods including k Nearest Neighbors (kNN), Artificial Neural Networks, weighted voting and support vector machines (SVM).

> SNL: *(cluster genes from microarray to build a model R for classification)*
> – Cluster $genes from $microarray-experiments
> – Classify $genes in $microarray-experiments

Computational Systems Biology
Systems biology is the study of systems of biological components including molecules, cells, organisms and entire species. Biological systems are dynamic and complex and their behavior may be hard to predict from the properties of individual parts. Computational systems biology employs quantitative measurements of the behavior of groups of interacting components based on bioinformatics and proteomics as well as mathematical and computational models that describe and predict dynamic behaviors (Kitano, 2002). Research on computational systems biology may be organized into four major areas: (1) system structure, such as gene regulatory and biochemical networks; (2) system dynamics, such as quantitative and qualitative analyses; (3) system control; and (4) sensing methods of a system. The SNL sentences for some representative problems are presented below:

> Gene regulatory networks
> - Simulate $gene-regulatory-network based on $time-course-microarray-data
> - Predict $gene-regulatory-network form $time-course-microarray-data

The problem of simulating a gene regulatory network may be solved using one or more of the following models:

1. Michaelis-Menten model (Bernot *et al.*, 2013)
2. S-System model (Voit, 2013)

3. Generalized mass action (Voit, 2013)
4. Lin-Log model (Visser and Heijnen, 2003)

The problem of predicting a gene regulatory network may be solved using one or more of the following models:

- S-System Parameter Estimation (Vilela *et al.*, 2008)
- GeneNet (Schäfer *et al.*, 2001)
- TimeDelay-ARACEN (Pietro Zoppoli *et al.*, 2010)

4.4.2 *Problem Space for Medical Intelligence*

Medical informatics is found at the intersection of healthcare. It is where skills in both medical and computer sciences come together in an effort to improve health care and patient outcomes. The goal of medical informatics is to help health care workers improve their way of working and the outcome of their performances. applications of medical informatics can contribute to better outcomes in medical care and decrease the costs of health care services through error reduction, providing patients with their needed information and supporting physicians with updated information and related knowledge().

As discussed, medical intelligence tools include OLAP and data mining that both allow an analyst to discover novel knowledge about the data stored in a medical data set. Specifically, OLAP allows an analyst to observe and interpret summary data. Data mining, on the other hand, can discover knowledge in a data set based on the process of extracting valid, previously unknown and potentially useful patterns and information from raw data. OLAP and data mining can be complementary: Analysts may navigate within a large dataset to form a specific dataset of interest using OLAP tools before discovering hidden patterns in the smaller dataset. Consequently they can be combined to enable analysts in obtaining data mining results from different portions of a data set and at different levels of generalization (Han, 1998).

In our study, we assume the analyst interacts with the MI system in two steps:

- Step 1: Compose one or more OLAP queries to extract from the dataset(s) patients of interest.
- Step 2: Compose one or more data mining queries to correlate the variables of interest or to compose one or more statistics analysis queries to estimate the variables of interest.

4.4.2.1 *Dimensions of Medical Data*

In this study, based on the OLAP theory, we allow an analyst to build a medical data cube. A data cube is a multidimensional database that is optimized for medical applications. In order to build a cube, we have to identify what to be analyzed in a medical dataset.

Figure 1 shows a medical database consisting of three tables: *PATIENT*, *DRUG* and *EXAMINATION*. They combined can form a data set consisting of the following "spaces" of variables (dimensions) for each patient, where each dimension consists of variables that we may not be interested in correlating their relationships:

- (MV) DEMOGRAPHIC VARIABLES: [D1] DOB, [D2] gender, [D3] race, [D4] (location, time)
- (BV) BIOLOGICAL AND ENVIRONMENT VARIABLES: [D5] risk factor, [D6] (exam, value), [D7] symptom, [D8] whether condition, [D9] time
- (DV) DIAGNOSIS VARIABLES: [D10] disease, [D11] time
- (TV) TREATMENT VARIABLES: [D12] (medication, dose, duration), [D13] (hospital, treatment), [D14] time
- (GV) GENOMIC VARIABLES: [D15] Nucleic Acids, [D16] time

We shall use the symbol MIV in the below to designate the union of MV, BV, DV, TV, and GV.

Each variable can be considered as a "dimension" of a multi-dimensional "cube", and a "cell" in the cube consists of a set of patient objects. At any instance of time we can focus on any sub-cube that may be derived.

Given a dataset, the analyst can use the following query pattern to form a "sub-cube":

GIVEN dataset(s)

Find patients conditions

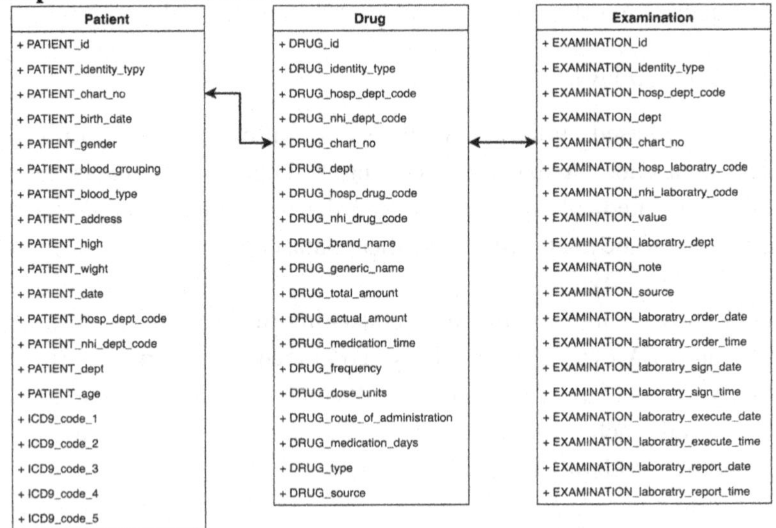

Figure 1. A medical database schema.

4.4.2.2 *Problem Space for Data Mining*

We may do a complete analysis of the problem space using different combinations of the variables from the five dimensions (MV, BV, DV, TV, GV) to describe problems in medical informatics. The SNL sentences that may be used to describe data mining problems are summarized in Table 1. The general SNL sentence pattern is:

Correlate $variable(s) in $space, $variable(s) in $space, …

In Table 1, we use a combinations calculator to find the number of possible combinations (PC) that can be obtained by taking a sub-set of items from a lager set:

$$C_n^k=\binom{n}{k}=\frac{P_n^k}{k!}=\frac{n!}{k!(n-k)!}\ \textit{for}\ C_n^k=\binom{n}{k}=\frac{P_n^k}{k!}=\frac{n!}{k!(n-k)!}$$

where *k* can be obtained from a larger set of n distinguishable objects where order does not count and repetitions are not allowed.

As can be seen form Table 1, we can identify 31 problems. Shown in Table 2, among the 31 problems, we have found 15 have been addressed in the literature, 10 have not yet be studied, and the other 6 do not make sense.

Table 1. Correlating variables in 1 to 5 spaces

Type	SNL	PC
1 Space	*Correlate $variable(s) in $space*	5
2 Spaces	*Correlate $variable(s) in $space and $variable(s) in $space*	10
3 Spaces	*Correlate $variable(s) in $space and $variable(s) in $space and $variable(s) in $space*	10
4 Spaces	*Correlate $variable(s) in $$space and $variable(s) in $space and $variable(s) in $space and $variable(s) $ in $space*	5
5 Spaces	*Correlate $variable(s) in $space and $variable(s) in $space and $variable(s) in $space and $variable(s) in $space and $variable(s) in $space*	1

*PC is Possible Combinations.

Table 2. Problems that have not been studied yet or do not make sense

Problems that have not yet been studied (10)
Correlate TV and GV
Correlate MV and TV and GV
Correlate DV and BV and TV
Correlate DV and BV and GV
Correlate DV and TV and GV
Correlate MV and DV and BV and TV
Correlate MV and DV and BV and GV
Correlate MV and DV and TV and GV

(*Continued*)

Table 2. (*Continued*)

Problems that have not yet been studied (10)
Correlate MV and BV and TV and GV
Correlate DV and BV and TV and GV
Problems that do not make sense (6)
Correlate MV and MV
Correlate BV and TV
Correlate BV and GV
Correlate MV and BV and TV
Correlate MV and BV and GV
Correlate BV and TV and GV

We present some case studies in the following.

Case Study – Correlate $variable in DV

Reference (Tai and Chiu, 2009) applies association rule mining to National Health Insurance (NHI) of Taiwan to explore the comorbidity of Attention Deficit Hyperactivity Disorder, and to examine the practicality of ARM (Associate Rule Mining) in the comorbidity studies using clinic databases.

Methods

Based on the clinical records of the enrollees of NHI, 18,321 youngsters aged 18 or less with a diagnosis of ADHD in 2001 were recruited to join the case group. All the clinical diagnoses made from 2000 to 2002 for those who were admitted to the medical center in 2008 and whose ICD-9 code id was 314 (Attention Deficit Hyperactivity Disorder) are extracted from the NHI database.

The following is the OLAP query:

GIVEN NHI

P1 = *Find patients whose disease include* [314] *and whose medical claims were between* 2000 and 2002

P2 = *Find patients whose disease include* [314] *and whose medical claims were between* 2002 and 2008

The following SNL sentence describes the problem of correlating the diseases diagnosed among the patients found by the OLAP query:

***Correlate disease in DV*[P1] *and disease in DV*[P2]**

The study identifies anxiety disorder, mild mental retardation and autism to be strongly related to ADHD.

Case Study – Correlate $variable in TV

Chinese herb medicine (CHM) is commonly used for Premenstrual Syndrome (PMS). Reference (Huang *et al.*, 2007) investigates the prescription patterns of CHM for PMS by using the NHIRD (National Health Insurance Record Database) in Taiwan.

Methods
Two million individuals between 1998 and 2011 randomly sampled from the NHIRD were extracted.

The following is the OLAP query:

GIVEN NHIRD
***Find patients whose disease includes* [625.4] *and whose medical claims were between* 1998 *and* 2011**

The SNL sentence that describes the problem of finding combinations of treatments for the patients found in the OLAP query is:

Correlate medication in TV

The study uses association rule mining (ARM) and social network analysis to explore the combinations of CHM treatments for PMS. It finds Jia-Wei-Xiao-Yao-San (JWXYS) had the highest prevalence (37.5% of all prescriptions) and also the core of the prescription network for PMS. For combination of two CHM, JWXYS with Cyprus rotundas L. are prescribed most frequently, 7.7% of all prescriptions, followed by JWXYS.

Case Study – Correlate $variable in BV and $variable in BV, Correlate $variable in BV and $variable in DV, Correlate $variable in TV and $variable in DV

Reference (Huang *et al.*, 2007) discusses how to decrease the threats is an important issue in medical treatment. This paper suggests to integrate data mining (DM) and case-based reasoning (CBR) to predict a chronic disease. The main processes include:

1) Adopting data mining to discover the rules from health examination data. The corresponding SNL sentence is "***Correlate $variable in BV and $variable in BV***".
2) Using data mining rules to predict a specific chronic disease. The corresponding SNL sentence is "***Correlate $variable in BV and $variable in DV***".
3) Using case based reasoning to support chronic disease diagnosis and treatments. The corresponding SNL sentence is "***Correlate $variable in TV and $variable in DV***"

This paper makes three critical contributions: (a) It suggests a systematical method of integrating DM techniques with CBR; (b) It shows that helpful implicit rules can be discovered from health examination data through DM techniques; and (c) It shows that the discovered rules from health examination data are helpful for chronic disease prognosis.

4.4.2.3 *Problem Space for Statistical Analysis*

The problem space for statistical analysis is shown in Table 3. As can be seen, we can identify 250 research problems. The general SNL sentence pattern is:

Estimate impact of $variable(s) in $space, ... on *$variable(s) in $space*

Table 3. Statistical analysis in 2 to 5 spaces

Type	SNL	PC
1 Space	*Estimate impact of $variable(s) in $space*	25
2 Spaces	*Estimate impact of $variable(s) in $space and $variable(s) in $space on $variable(s) in $space*	50
3 Spaces	*Estimate impact of $variable(s) in $$space, $variable(s) in $space and $variable(s) in $space on $variable(s) in $space*	50
4 Spaces	*Estimate impact of $variable(s) in $space, $variable(s) in $space, $variable(s) in $space and $variable(s) in $space on $variable(s) in $space*	25
5 Spaces	*Estimate impact of $variable(s) in $space, $variable(s) in $space, $variable(s) in $space, $variable(s) in $space and $variable(s) in $space on $variable(s) in $space*	5

Case Study - Estimate impact of $variables in DV on $variables in DV

Reference (H.-C. Wu *et al.*, 2013) addresses the incidence and relative risk of stroke among patients with biopolar disorder on a seven year follow up study. This paper aims to estimate the incidence and relative risk of stroke and post stroke that cause all-cause mortality among patients with biopolar disorder.

Methods

The authors identified a study population from the NHI Research Database (NHIRD) in Taiwan database between 1999 and 2003 that includes 16821 patients with bipolar disorder and 6728 age and sex matched control participants without bipolar disorder. The incidence of ICD9 code is between 430 and 438, patient survival rate after stroke are calculated for both groups using data from the NIHRD between 2004 and 2010.

The following OLAP queries build the target datasets:

GIVEN NHIRD
***P11 = Find patients whose disease includes* [430:438] *and whose medical claims were between* 2004 *and* 2010**

GIVEN P11
***P12 = Find patients whose disease includes* [434.91] *and whose medical claims were between* 2004 *and* 2010**

GIVEN NHIRD
***P21 = Find patients whose disease does not include* [430:438] *and whose medical claims were between* 2004 *and* 2010**

GIVEN P21
***P22 = Find patients whose disease includes* [434.91] *and whose medical claims were between* 2004 *and* 2010**

The SNL sentence that describes the problem of finding combinations of treatments for the patients found in the OLAP query is:

Estimate impact of disease (bipolar) in DV (P11,P12) on disease (stroke) in DV (P21, P22)

The study uses a proportional-hazards model to compare the seven-year stroke-free survival rate and all-cause mortality rate across the two cohorts after adjusting for confounding risk factors. It uses Relative Risk (RR), Incidence Rate (IR) and confidence interval (CI) to investigate the Incidence and Relative Risk of stroke among patients with bipolar disorder.

Case Study – Estimated impact of $variables in DV on $variables in DV

Reference (Wang *et al.*, 2014) evaluates the risk of depressive disorders among patients with rheumatoid arthritis (RA, ICD-9 is 714). The study uses data of 18285 patients from NHIRD in Taiwan. Patients are observed for a maximum of 10 years to determine the rates of newly diagnosed depressive disorders, and the Cox regression method is used to identify the risk factors associated with depressive disorders in RA.

The following OLAP query constructs the target dataset:

GIVEN NHIRD
Find patients whose disease includes* [714] *and whose medical claims cover 10 years

The SNL sentence that describes the problem of estimating the impact of one disease **(rheumatoid arthritis)** on another **(depressive disorders**) for the patients found in the OLAP query is:

***Estimated impact of disease (*rheumatoid arthritis*) in DV on disease in DV* (depressive disorders)**

Case Study – Estimated impact of air pollution in BV on disease in DV

Reference (Chang *et al.*, 2014) discusses that the associations between the levels of nitrogen dioxide (NO_2) and carbon monoxide (CO) exposures and dementia remains poorly. It obtains data of 29547 people form NHIRD Taiwan including data on 1720 patients diagnosed with dementia between 2000 and 2010. The incidence of ICD9 code is 294.20.

The following OLAP query establishes the target dataset:

GIVEN NHIRD and air pollution data
***Find patients whose disease includes* [294.20] *and whose medical claims were between* 2000 *and* 2010**
***Find and air pollution data between* 2010 *and* 2010**

The SNL sentence that describes the problem of estimating the impact of environment ***(air pollution)*** on a disease **(dementia)** is:

Estimate impact of weather condition (air pollution) in BV on disease (dementia) in DV

The paper evaluates the risk of dementia among four levels of air pollutants. It uses Relative Risk (RR) and confidence interval (CI) to investigate the Incidence and Relative Risk of dementia among patients to discover the associations between the levels of nitrogen dioxide (NO2) and carbon monoxide (CO) exposure and dementia.

Case Study – Estimated impact of patient self-reports in MV and medication in TV on disease in DV

Reference (C.-S. Wu *et al.*, 2014) addresses the concordance between claims records in the NHIRD Taiwan and patients' self-reports on clinical diagnoses, medication use, and health system utilization. It uses the data of 15574 participants collected from 2005's Taiwan National Health Interview Survey (NHIS). The ICD-9 codes used are 401-405, 250, 272, 430-438, 293, 580-593, 393-398, 402, 404, 410-414, 426, 427, 274, 490-492, 494, 496, 733, 571, 070, 140-155, 162, 164, 172-175, 180, 182, 183, 185, 188, 200-208, 296, 300, 311, 714, 715. It uses positive agreement, negative agreement and Cohen's kappa statistics to examine the concordance between claims records and patient self-reports.

The following is the OLAP query:

GIVEN Taiwan National Health Interview Survey (NHIS)
***Find patients whose disease includes* [401-405, 250, 272, 430-438, 293, 580-593, 393-398, 402, 404, 410-414, 426, 427, 274, 490-492, 494, 496, 733, 571, 070, 140-155, 162, 164, 172-175, 180, 182, 183, 185, 188, 200-208, 296, 300, 311, 714, 715] *and whose medical claims in* 2005**

The SNL sentence that describes the problem of finding the concordance of patients' self-reports on clinical diagnoses, medication use and health system utilization for the patients found in the OLAP query is:

Estimate impact of patient self-reports in MV and medication in TV on disease in DV

Case Study – Estimate impact of $variable in GV on $variable in DV

Reference (Lee *et al.*, 2011) proposes that a database of gene environment interactions pertains to blood lipid traits, cardiovascular disease and type 2 diabetes. This database is accorded a more prominent role in modifying the relationship between genetic variants and clinical measures of diseases, the consideration of gene-environment interactions is a must. From the literature a database of GxE interactions relevant to nutrition, blood lipids, cardiovascular disease and type 2 diabetes over 550 such interactions are incorporated into a single database, along with over 1430 instances.

OLAP Query:

GIVEN Gene-Environment Interactions Pertaining
***Find patients whose diagnoses include* [blood lipid traits, cardiovascular disease, type 2 diabetes]**

The following SNL sentence describes the problem of correlating genomic variables among the patients found in the OLAP query:

Estimate impact of genomics GV on disease in DV

4.5 Conclusions

In this chapter, we survey different bioinformatics tools, data mining and OLAP tools, and statistical analysis tools for predication analysis and decision support for biomedical applications. We describe a Structured Natural Language (SNL) that can express many problems in BMI with a finite number of sentence patterns. We show how OLAP, data mining and statistical analysis tools may be linked to solve problems in computational medicine with a uniform interface, and how a language-based approach may help in discovering new problems.

This study shows an attempt to summarize some biological and medical informatics problems with a finite number of sentence patterns to simplify the interface between human and computer. As shown for many sentences we are not able to identify supporting references due to our limited knowledge. It is likely that there exist other sentence patterns of interest, and it is our wish that the vocabulary can be incrementally expanded to cover most, if not all, problems in computational biology and medicine.

One problem that is not addressed by this paper is how to solve a specific problem automatically based on the existing tools. For each problem sentence, if applicable we give a case study that shows an instance of the corresponding problem has been solved. For different instances of a problem we may need to apply variations of the tools employed in the case study. In addition, we need to connect and convert the methods used in a case study into an algorithm that is suitable for automation and parameterization. What is described in the paper is merely the beginning of a long-term effort to provide an integrated platform for biomedical problem solving.

Acknowledgment

This work of CCNW, PCYS and JJPT are supported in part under grant numbers NSC 102-2632-E-468-001-MY3 and MOST 105-2632-E46-002 from the Ministry of Science and Technology, Taiwan and Asia University. The views, opinions and/or findings contained in this report are those of the authors and should not be construed as an official National Science Council position, policy or decision unless so designated by other documentation.

References

Abu-Jamous, B. *et al.* (2013) Paradigm of Tunable Clustering Using Binarization of Consensus Partition Matrices (Bi-CoPaM) for Gene Discovery. *PLOS ONE*, **8**, e56432.

Altschul, S.F. *et al.* (1990) Basic local alignment search tool. *Journal of Molecular Biology*, **215**, 403–410.

American Medical Informatics Association American Medical Informatics Association. *httpswww.amia.orginside*.

Bernot, G. *et al.* (2013) Modeling and Analysis of Gene Regulatory Networks. In, *Modeling in Computational Biology and Biomedicine*. Springer Berlin Heidelberg, Berlin, Heidelberg, pp. 47–80.

Boulesteix, A.-L. and Strimmer, K. (2007) Partial least squares: a versatile tool for the analysis of high-dimensional genomic data. *Brief Bioinform*, **8**, 32–44.

Chang, K.-H. *et al.* (2014) Increased Risk of Dementia in Patients Exposed to Nitrogen Dioxide and Carbon Monoxide: A Population-Based Retrospective Cohort Study. *PLOS ONE*, **9**, e103078.

Han, J., Kamber, M., Pei, J. (2011) Data Mining: Concepts and Techniques.

Fryer, R.M. *et al.* (2002) Global Analysis of Gene Expression: Methods, Interpretation, and Pitfalls. *Nephron Exp Nephrol*, **10**, 64–74.

Hall, M. A. (2000) Correlation-based feature selection of discrete and numeric class machine learning. In Proceedings of 17th Int'l Conf. Machine Learning, 359-366.

Han, J. (1998) OLAP Mining: An Integration of OLAP with Data Mining. In, *Data Mining and Reverse Engineering*. Springer US, Boston, MA, pp. 3–20.

Hristovski, D., Dinevski, D., Kastrin, A., and Rindflesch, T.C. (2015a) Biomedical question answering using semantic relations. *BMC Bioinformatics 2009 10:1*, **16**, 1.

Hristovski, D., Dinevski, D., Kastrin, A., and Rindflesch, T.C. (2015b) Biomedical question answering using semantic relations. *BMC Bioinformatics 2009 10:1*, **16**, 6.

Huang, M.-J. *et al.* (2007) Integrating data mining with case-based reasoning for chronic diseases prognosis and diagnosis. *Expert Systems with Applications*, **32**, 856–867.

Ikeda, S. *et al.* (2011) A MODEL FOR OBJECT RELATIONAL OLAP. *International Journal on Artificial Intelligence Tools*, **19**, 551–595.

Kim, S.Y. *et al.* (2006) Comparison of various statistical methods for identifying differential gene expression in replicated microarray data. *Stat Methods Med Res*, **15**, 3–20.

Kitano, H. (2002) Computational systems biology. *Nature*, **420**, 206–210.

Lander, E.S. *et al.* (2001) Initial sequencing and analysis of the human genome. *Nature*, **409**, 860–921.

Lee, Y.-C. *et al.* (2011) A Database of Gene-Environment Interactions Pertaining to Blood Lipid Traits, Cardiovascular Disease and Type 2 Diabetes. *Journal of data mining in genomics & proteomics*, **2**.

Lowe, H. J. and Barnett, G. O. (1994) Understanding and using the medical subject headings (MeSH) vocabulary to perform literature searches. *JAMA*, **14**, 1103–1108.

MACGREGoR P. F. and SqUIRE J. A. (2002). Application of microarrays to the analysis of gene expression in cancer. *Clinical Chemistry*, **48**, 1170–1177.

Maojo,V. and Kulikowski, C.A. (2003) Bioinformatics and Medical Informatics: Collaborations on the Road to Genomic Medicine? *Journal of the American Medical Informatics Association*, **10**, 515–522.

Mutch, D.M. *et al.* (2001) Microarray data analysis: a practical approach for selecting differentially expressed genes. *Genome Biol.*, **2**, preprint0009.1.

Natarajan, J. (2013) Text Mining Perspectives in Microarray Data Mining, *ISRN Computational Biology*, **2013**, 5.

National Library of Medicine National Library of Medicine. *httpswww.nlm.nih.gov*.

Pietro Zoppoli *et al.* (2010) TimeDelay-ARACNE: Reverse engineering of gene networks from time-course data by an information theoretic approach. *BMC Bioinformatics 2009 10:1*, **11**, 154.

Rud, O.P. ed. (2012) Business Intelligence Success Factors John Wiley & Sons, Inc., Hoboken, NJ, USA.

Schäfer, J. *et al.* (2001) Reverse engineering genetic networks using the GeneNet package. *J Am Stat Assoc.*

Sheu, P.C.-Y. and T. Kitazawa. (2007) From Semantic objects to Semantic Software Engineering. *International Journal of Semantic Computing*, **01**, 18.

Sheu, P.C.-Y. (2007) Semantic Computing. *International Journal of Semantic Computing*, **01**, 9.

Svrakic, N.M. *et al.* (2003) Statistical approach to DNA chip analysis. *Recent Prog Horm Res*, **58**,75–93.

Tai, Y.-M. and Chiu, H.-W. (2009) Comorbidity study of ADHD: Applying association rule mining (ARM) to National Health Insurance Database of Taiwan. *International Journal of Medical Informatics*, **78**, e75–e83.

Tarczy-Hornoch, P. and Minie, M. (2005) Bioinformatics Challenges and Opportunities. In, *Medical Informatics*, Integrated Series in Information Systems. Springer US, Boston, pp. 63–94.

Tusher, V.G. *et al.* (2001) Significance analysis of microarrays applied to the ionizing radiation response. *PNAS*, **98**, 5116–5121.

Vilela, M. *et al.* (2008) Parameter optimization in S-system models. *BMC Systems Biology 2008 2:1*, **2**, 35.

Visser, D. and Heijnen, J.J. (2003) Dynamic simulation and metabolic re-design of a branched pathway using linlog kinetics. *Metabolic Engineering*, **5**, 164–176.

Voit, E. O. (2013) Biochemical Systems Theory: A Review, *ISRN Biomathematics*, **2013**, 53.

Wang, S.-L. *et al.* (2014) Risk of Developing Depressive Disorders following Rheumatoid Arthritis: A Nationwide Population-Based Study. *PLOS ONE*, **9**, e107791.

Wu, C.-S. *et al.* (2014) Concordance between Patient Self-Reports and Claims Data on Clinical Diagnoses, Medication Use, and Health System Utilization in Taiwan. *PLOS ONE*, **9**, e112257.

Wu, H.-C. *et al.* (2013) The Incidence and Relative Risk of Stroke among Patients with Bipolar Disorder: A Seven-Year Follow-Up Study. *PLOS ONE*, **8**, e73037.

Zhang, R. *et al.* (2014) Using semantic predications to uncover drug–drug interactions in clinical data. *Journal of Biomedical Informatics*, **49**, 134–147.

Chapter 5

Investigating interactions between proteins and nucleic acids by computational approaches

Wen-Pin Hu[a], Hui-Ting Lin[b], Jeffrey J. P. Tsai[a] and Wen-Yih Chen[c]

[a]*Department of Bioinformatics and Medical Engineering, Asia University, Taichung 41354, Taiwan*

[b]*Department of Physical Therapy, I-Shou University, Kaohsiung 82445, Taiwan*

[c]*Department of Chemical and Materials Engineering, National Central University, Jhong-Li 32001, Taiwan*

5.1 Introduction

Molecular simulations are wildly used in studying many types of biomolecular interactions, such as protein-protein interaction, drug-receptor interaction, protein-nucleic acid (NA) interaction, ion transport, etc. The first molecular dynamics (MD) simulation was reported by Alder and Wainwright in the late 1950s [1, 2]. In the early 1960s, Rahman performed molecular dynamics with a practical potential for a system of 864 particles in liquid argon [3]. Until the late-1970s, MD simulations were carried out to study proteins [4, 5]. Since the biological systems are very complex, suitable computation algorithms and powerful computers are indispensable for solving the problems in the life sciences. With the revolutionary advances in computers and the improvements on algorithms over more than half a century, molecular simulations have become a valuable research tool to obtain abundant atomic-level information in the fields of biology, chemistry and physics. Besides, accurate and reasonable models of molecular structures are also very important for exploring molecular interactions by using computational approaches. Many researchers utilized experimental techniques, like X-ray diffraction or nuclear magnetic resonance (NMR), to determine the structures of biomolecules. Biological models generated by experiments were usually available for download from the Protein Data Bank (PDB). However, three-dimensional structures of many proteins are not determined yet from experimental techniques. For this reason, bioinformatics scientists developed different algorithms and provided web server services to predict and analyze protein structure, function and even to build 3D models, such as SWISS-MODEL and Phyre2 [6–9].

Compared with the simulation studies of proteins, simulations of nucleic acids (NAs) basically fall behind protein simulations. The first simulation study of NA was appeared until 1983 [10, 11]. Until new force fields was proposed, the number of studies related to the simulation of NAs increased gradually after the mid-1990s, and the long-range electrostatic effects were proper represented and taken into the calculations in new force fields [12]. Particle mesh Ewald method (PME) is particularly central to the simulations of nucleic acids because NAs have negatively charged backbones and structures are usually entirely non-globular. The PME uses a Fast Fourier Transform (FFT) to calculate the long range interactions and then the charges of NAs are mapped onto grid positions. For performing MD simulations, MD simulation programs based on the CHARMM force field are most frequently used. The versions of CHARMM force fields include CHARMM19, CHARMM22, CHARMM27, and CHARMM36. CHARMM27 and CHARMM36 are improved force fields for simulating DNA, RNA, and lipids. CHARMM19, CHARMM22 and CHARMM36 are suitable for the simulations of proteins. Popular simulation packages include CHARMM, GROMOS, GROMACS, NAMD and the Assisted Model Building with Energy Refinement (AMBER) that have been developed specifically for simulations of biomolecules and each program generally adopts different methods in the evaluation of biomolecular interaction. AMBER and

CHARMM are the mostly used force fields for the implementation of molecular simulations. Weiner *et al.* [13] developed the AMBER force field that was original for the calculations of proteins and NAs. Nowadays, several types of AMBER forces fields with improved parameters (ff94, ff96, ff98, ff99, ff99SB, etc.) designed for the simulations of proteins, peptides and NAs. Some modified AMBER force fields, parm94, parm99 and parmbsc0, show impressive performances in modeling a large number of DNA structures [12, 14, 15].

Interactions between proteins and nucleic acids play important roles in many biological activities, which involve in degradation of nucleic acids, protein synthesis, DNA replication, RNA transcription, and RNA splicing [16]. In the past, many three-dimensional structures of NAs are unavailable, which is one of the limitations for simulating protein-NA interaction. Currently, some web servers provide the functions for predicting and generating the 3D-structural models of NAs [17–19]. These advances make investigating protein-NA interactions via computational approaches more possible. Computational simulations indeed can help interpreting protein–nucleic acid interactions and complementing experimental results. Therefore, simulation studies of protein-NA complexes attracted great attention from many scientists due to the capability of characterizing the binding domain of protein for NA and visualizing the interaction forces between the protein and NA.

This chapter focuses on the computer simulation studies of protein-aptamer simulations. Aptamers are short single-stranded DNA or RNA, and they can form a special stem-loop secondary structure and have the specificity for recognizing target molecules. The target molecules can be viruses, cells, proteins, ions, drugs, toxins, peptides and bacteria. First, we present the interacting forces between proteins and nucleic acids. Second, we describe the two most widely used force fields for simulations briefly; AMBER and CHARMM. Available modeling tools of proteins and aptamers, and the experimental procedures and computational approaches for selecting aptamers are introduced in the chapter.

5.2 The forces between proteins and nucleic acids

The bindings between proteins and nucleic acids can be classified as specific or non-specific interactions. For instance, theα-helix motif in the protein usually dominates the specific-DNA interactions through hydrogen bonds and ionic interactions occurred within the motif of protein and the major groove of the DNA. In general speaking, there are five major forces that occur when proteins and nucleic acids interact with each other. Besides, nucleic acid sequences also can fold into special secondary and tertiary structures to recognize and bind different protein targets by the additional mechanism of shape complementarity. Here, the five major types of forces, ionic interactions, hydrogen bonds, van der Waals, hydrophobic and base stacking forces, are descripted as follows.

5.2.1 *Ionic interactions*

Ionic interactions are generated from the electrostatic interactions between groups of opposite charge. The salt concentration can influence the strength of Ionic interactions between two molecules. In the solution with high salt concentration, the week ionic interactions lead to destabilize structures of DNA-protein complexes. Because of the screening effect generated in the high-salt solution, the electrostatic attraction force becomes negligible. Electrostatic interactions are long-ranged forces, which dominate the protein-nucleic acid and protein-protein binding. The charged residues on the protein (often positively charged) can interact with the atoms of opposite charge (negatively charged) in the nucleic acid. Actually, the overall ionic interactions between molecules include attraction and repulsion forces. Electrostatic forces have the greatest contribution to the energy of hydrogen bond. For protein–protein interfaces, hydrogen bonds and salt bridges are generated abundantly on protein surfaces. The salt bridges are also considered as a special form of hydrogen bonds [20]. In an earlier study, Xu *et al.* [21] reported that electrostatic interactions have a more significant role in binding than in folding. Besides, they found that interfacial hydrogen bonds and salt bridges, as the major contributors to the electrostatic interactions between proteins. In protein–DNA and protein–RNA recognition, the significance of electrostatic interactions was also verified by many studies [22–24].

5.2.2 *Hydrogen bonds*

Hydrogen bonding is a kind of intermolecular attraction forces, which occurs between polar groups of different molecules. Basically, the hydrogen bond is categorized as a type of weak chemical bond. The hydrogen bond is not only described special type of dipole-dipole attraction but also it has a few features of covalent bonding. The formation of hydrogen bond arises from a hydrogen atom covalently bound to a highly negative charged atom (e.g., nitrogen (N), oxygen (O) or fluorine (F) atom) and another atom carried a negative charge. The hydrogen bond is usually expressed by the formulation X-H...Y, where the dots represent the hydrogen bond. Comparing with the covalent and ionic bonds, the hydrogen bond is weak. On the other hand, the hydrogen bond is stronger than the van der Waals forces. The strength of hydrogen bond can be influenced by temperature, pressure, bond angle, and environment [25]. Therefore, the strength of hydrogen bond can range from 4 kJ to even >40 kJ per mole of hydrogen bonds [26]. In the structure of DNA, the hydrogen bonds form to link bases and have the strength of 8-12 kJ/mol [27]. Except for inorganic molecules, this type of bond commonly forms in the biological molecules such as nucleic acids and proteins.

5.2.3 *van der Waals forces*

Van der Waals force can be classified as three types: dipole-dipole interaction (Keesom force), dipole-induced dipole interaction (Debye force) and London dispersion force. Unlike chemical bonding, van der Waals forces are very weak and the forces originate from the distribution of electronic charge around the molecules. The electrons of molecules are always in motion especially when the charge distribution does not reach to a stable state. Like other intermolecular forces, van der Waals forces include the attractive and repulsive electrical forces between atoms and molecules. The attractive electrical force between two atoms can make them come closer to each other and two atoms are finally separated by the van der Waals contact distance. If two atoms come too close to each other, the repulsive force becomes the dominant force. Because the outer electron cloud of an atom overlaps that of another atom causes the repulsive force. Van der Waals interaction contribute the strength of bond from 2 to 4 kJ/mol per atom pair.

5.2.4 *Hydrophobic force*

The hydrophobic force is important for biological molecules such as proteins and nucleic acids, which influences the behavior of water at the interface. Hydrophobic forces are short range and very sensitive to surface structure of molecule. The structure of proteins basically have hydrophobic and hydrophilic amino acids, and the hydrophobic amino acids fold to form a hydrophobic core as the protein dissolves in the solution. At the same time, the hydrophilic side chains of amino acids expose to the water, and this consequence stabilizes the protein in the folded state. For the structure of DNA double helix, van der Waals forces and hydrogen bonds generate between complementary base pairs, and hydrophobic forces occur between nitrogenous bases and the surrounding water sheath. These four main forces contribute to stabilizing the native DNA structure.

5.2.5 *Base stacking force*

Stacking between adjacent bases is also a key factor that is responsible for the stability of the DNA double helix [28]. Stacking interactions take place between complementary base pairs of double-stranded DNA and depend on the dipole moments and the aromaticity of the bases. The base stacking force is short ranged and can be characterized by an attraction potential and a strong repulsion potential [29]. The strength for the stacks of G-C base pairs is stronger than that for the stacks of A-T base pairs. For dsDNA, base staking forces are very central in maintaining the structure. Unlike the function in the dsDNA, the base staking forces can help ssDNA bind with proteins because bases are usually bound by stacking with aromatic protein side chains [30]. Base stacking forces depend on several noncovalent forces, and hydrophobic and electrostatic interactions are the main forces. In the simulation model, base-stacking interaction is considered as the main factor determining the high extensibility and unwinding instability of DNA [30, 32]. Nucleotide base pairs can produce base-stacking energy, which primarily generates from the noncovalent van der Waals interactions between adjacent base pairs [32]. The noncovalent van der Waals interactions has the function of stabilizing the stacking orientation, besides electrostatic effects of interacting dipoles in the structure also absolutely affect the stability of stacking.

5.3 Simulations on the interactions between proteins and nucleic acids

5.3.1 *Force fields*

5.3.1.1 *AMBER force field*

The term "AMBER force field" generally refers to a family of force fields for molecular dynamics of biomolecules, which is originally developed by Peter Kollman's group at the University of California, San Francisco. The force field is developed for the simulation of macromolecules, and it is necessary to set appropriate parameters of the force field (e.g., bonds, angles, dihedrals, and atom types in the system). Many standard sets of parameters exist and provide in the simulation programs. For the simulations of proteins and nucleic acids, the *ff14SB* AMBER force field is suitable. The *ff14SB* force field is an improved version of the *ff99SB* AMBER force field, originally developed by Hornak *et al.* in 2006 [33]. The *ff14SB* force field runs simulations well in the system combined with protein, nucleic acid and water models. According to the reference manual of AMBER, *OL15* and *OL3* are more specific force fields for the simulations of DNA and RNA molecules, respectively. For example, Krepl *et al.* performed MD simulations of protein/RNA complexes and they adopted the $ff99bsc0\chi_{OL3}$ force field for RNA and *ff14SB*, *ff12SB* and *ff99SB* force field for proteins [34].

5.3.1.2 *CHARMM force field*

The CHARMM Force Field is a frequently used force field in the study of biomolecules, original developed by Karplus and co-workers in 1983 [35]. Currently, there are several versions of CHARMM Force Field available for applying in the simulations of different biological systems, including proteins, peptides, nucleic acids, small molecule ligands, lipids, prosthetic groups and carbohydrates. CHARMM program is continuously maintained by a large group of developers led by Martin Karplus, and free CHARMM program is also available for download on the website (https://www.charmm.org/charmm/). The CHARMM19 adopts a united atom force field where only hydrogen atoms belonging to polar groups are explicitly included [36]. Continuous improvements and development efforts in the CHARMM additive force field give more accurate simulations of different biomolecules. Nowadays, there are several versions of the additive force field. CHARMM22, CHARMM27, and CHARMM36 are all belonged to atom force fields. CHARMM22 [37] is suitable for the simulations of proteins, and CHARMM27 [38] is designed for nucleic acids (DNA and RNA). CHARMM36 [39], CHARMM37 [40, 41] and CGenFF [42] are developed with the optimal simulation parameters for lipids, carbohydrates and drug-like molecules (small molecules), respectively. CHARMM [37] and AMBER [14] are the commonly used two force fields to simulate nucleic acid-protein complexes [43]. Both force fields can apply in the DNA and RNA models. Except for the CHARMM program, other MD programs like AMBER, GROMACS and NAMD also adopt the CHARMM additive force field in MD simulations. HyperChem, a molecular modeling software, incorporate the CHARMM force field and name it as Bio+ force field.

5.3.2 *Sources of molecular models*

5.3.2.1 *Protein models*

The Protein Data Bank offers 120,388 biological macromolecular structures that can be accessible in July 2016 and the total number of macromolecular structures is still increasing. The 3D structural models of proteins deposited in the database are generated by X-ray crystallography and NMR. The experimental process for generating a 3D protein model is time-consuming and have a great chance of failure. Nevertheless, there still has a large number of known protein sequences without the 3D protein models. SWISS-MODEL is an automated and web-based protein homology-modeling server, which is the most popular used homology-modeling server [44]. The homology-modeling approaches contains four essential steps, including template selection, alignment, model building and model assessment. These steps are iterative repeated until no further improvement can be found in the model structure. ModBase [45], the Homology Modeling Automatically (HOMA) web site [46], I-TASSER server [47], MODELLER [48] and others are all useful ways to create homology models. These computational methods of homology modeling give solutions to build the protein models of target protein sequences by using the data of homologous proteins with known 3D structures solved in experiments. These techniques make detailed analysis of protein structure and function more possible, especially these protein models also bring significant contribution to molecular simulations.

5.3.2.2 *Models of Nucleic acids*

Although NMR and X-ray crystallography are commonly used experimental techniques to obtain the 3D structures of biomolecules, available structural models of nucleic acids are relatively much less than that of the proteins. Some 3D shapes of nucleic acids and protein-nucleic acid complexes are available in the Protein Data Bank. For the prediction of nucleic acid structure, many researchers develop computational methods such as structure prediction software tools for nucleic acids [17–19, 49]. Most of these tools [49, 50] and some webservers, like the RNAfold [51] and RNAstructure [52], provide the functions of predicting secondary structures of single stranded RNA or DNA sequences. The RNAstructure webserver combines four separate analysis and prediction algorithms: calculating a partition function, predicting a maximum free energy (MFE) structure, finding structures with maximum expected accuracy, and pseudoknot prediction [52]. Users can upload a sequence file or type the sequence (DNA or RNA) into the input box, and then submit the query. After calculating, partition, ProbKnot, MaxExpect and fold results are expressed with figures in the browser. However, the 3D structures of nucleic acids are still unavailable from these analyses. For the RNAfold webserver, the server provides the minimum free energy and thermodynamic ensemble predictions of the secondary structures of nucleic acids with graphs and dot-bracket notations. By using the dot-bracket notation and sequence information, the RNAComposer webserver can generate the prediction of RNA 3D structures automatically. The method used by the webserver is based on the machine translation principle, and it operates on the RNA FRABASE database [17, 53]. Actually, the RNAComposer webserver also can complete the whole processes by just giving the RNA sequence only. Because the webserver can obtain the information of dot-bracket notation from the RNAfold program. Some software also provide the functions of drawing or build the 3D structures of nucleic acids, like HyperChem, Discovery Studio, ChemDraw, etc. The 3D-DART web server can generate 3D structural model of DNA from a sequence according to manual selection of structural parameters: (1) nucleic acid type (A-form or B-form); (2) global and local bend-angle; (3) orientation of the bend-angle in Euler-angle space; (4) location of the bend-angle in the sequence; (5) custom values for base-pair and base-pair step parameters [18].

5.3.3 *Simulations of protein-nucleic acid interactions*

Bini *et al.* [54] use computational approach for the selection of thrombin-binding aptamers (TBA). Aptamers are usually selected from experimental procedures called as the systematic evolution of ligands by exponential enrichment procedure (SELEX). Before the onset of SELEX, random nucleic acids are generated as the nucleic acid libraries. Basically, a library generally contains $10^{14}{\sim}10^{15}$ different sequences. There are four main steps in the cycle of SELEX: (1) incubation with target protein; (2) separation of unbound nucleic acids and conservation of protein-nucleic acid complexes; (3) elution of nucleic acids (aptamers); (4) amplification by PCR. The cycle needs to repeat 8-16 times and then the sequence of selected aptamer that can bind to the target protein with high specificity needs to be sequenced. In the study of Bini's group [54], they choose a well-known sequence of 15-mer thrombin binding aptamer and generate 994 mutated DNA sequences based on the aptamer. They use the protein-ligand docking program in OpenEye Scientific Software to run simulations. The simulation software utilize the Merck molecular forcefield (MMFF94) that is developed by Halgren Thomas [55] and is a reparameterized CHARMM force field. Because the characteristics of thrombin have been thoroughly studied by many researchers. Thus, they choose the TBA as a model and try to study the binding of thrombin and TBA by simulations. Their study indicated that there is a good correlation between simulation and experimental results, and they also screen an aptamer has better performance in binding with the thrombin from the mutated DNA sequences.

In our group, we have been tried to use the commercial simulation software, Discovery Studio, to study the protein-nucleic acid interactions in the CHARMM force field. In the beginning, we adopted the well-known sequence of TBA and three mutated oligonucleotides reported by the Bini's group [54] and used the ZDOCK algorithm to calculate the protein-nucleic acid interaction. The ZDOCK algorithm adopts a Fast Fourier Transform (FFT)-based docking algorithm and uses a simple shape complementarity method called PSC for identifying docking conformations [56]. Desolvation (DE) and electrostatic (ELEC) energy terms are also combined in the PSC method to rank the docked poses. Actually, the ZDOCK algorithm is developed for the docking prediction of protein-protein complexes. In the calculation, the receptor and ligand are treated as rigid bodies, but the scoring functions are also called as soft docking functions because six rotational and translational degrees of freedom are fully discovered in the docking. For the TBA-binding aptamers with short sequences (15-mer), the results of our study indicate that the ZDOCK method is able to be used in the computational simulation of protein-aptamer interaction [57]. A representative simulation result is presented in Figure 1. Since our simulation results are consistent with simulation and experimental results reported by Bini's group. According to these results, we even think that the ZDOCK method has the potential of using in the computational selection of target-specific aptamers.

However, the performance of using the ZDOCK score for evaluating the interactions between the protein and the long-sequence aptamers is not as expected in our another study [58]. In this study, we choose 15 RNA aptamers that have been demonstrated their binding ability to the angiopoietin-2 (Ang2) from literatures [59, 60]. These RNA aptamers are 40 or 41 mers in length. In the light of ZDOCK score, the prediction results of protein-aptamer interactions are not consistent with previous reports [59, 60]. Therefore, we use the ZRANK algorithm to rescore the docking predictions obtained from ZDOCK. The ZRANK scoring function takes van der Waals attractive and repulsive energies, short and long range repulsive and attractive energies, and desolvation terms into calculation. This ZRANK program with an optimized energy function can significantly improve the success rate of prediction from the initial ZDOCK. By using the ZRANK score, we find that the results of docking prediction are consistent with the findings from previous reports [59, 60]. Based on this study, we conclude that the ZRANK scoring function can give more accurate predictions of protein-aptamer interactions when these aptamers have longer sequences.

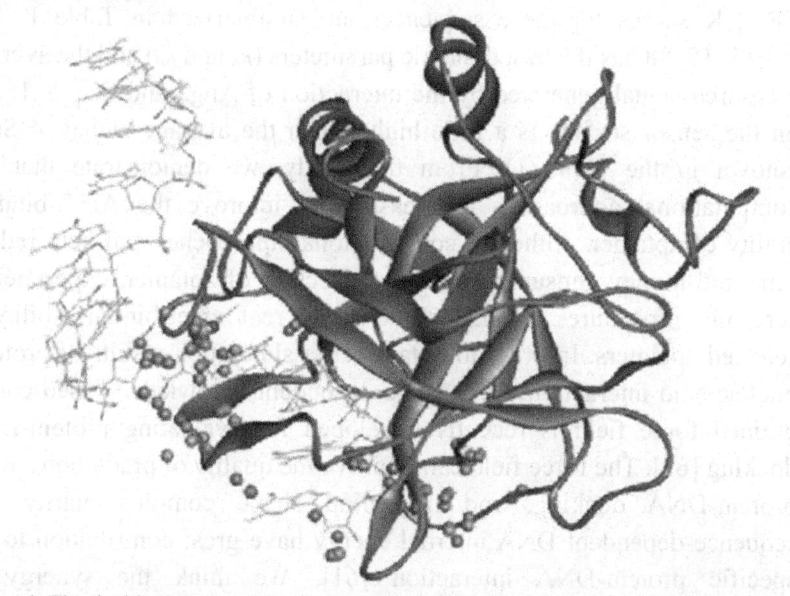

Figure 1. The docking results of thrombin (shown with ribbons) and the sequence of Best TBA reported by Bini *et al.* [54] by using the ZDOCK algorithm. The image shows the best docking result of the two molecules. The dots with different colors represents different docking poses. The color of dot approximates to red (darkest in this gray image) that means it is a better docking pose.

In the next study [58], we select 3 RNA sequences out of the 15 Ang2-binding aptamers reported in previous reports [59, 60] as the parent sequences for generating two-point mutant sequences. A total of 189 mutant sequences are created, and these sequences are modeled to build 3-dimensional structures by using the RNAComposer webserver [19]. The binding affinity of these sequences to Ang2 is calculated with the ZDOCK and ZRANK algorithms. Finally, three mutant sequences are selected in the end of computational selection. Except for the computational simulations, we also carry out experiments on the surface plasmon resonance (SPR) biosensor to ascertain the real binding reactions between aptamers and Ang2. Aptamers are immobilized on the sensor surface through the biotin /streptavidin interaction, and the buffer solution containing 0.5 μM Ang2 is flowed over the sensor surface to test the interaction of aptamer and Ang2. Figure 2 shows the experimental results. Seq1 and Seq16 are the known sequences with the best and worst binding affinity to Ang2, respectively [59, 60]. The experimental data and the ZRANK scores for these sequences are summarized in Table 1. The Seq15_15_38 has the best dynamic parameters (k_a and k_d) and the average measured signal generated by the interaction of Ang2 and Seq15_15_38 on the sensor surface is a little higher than the average signal of Seq1 (shown in the Table 1). From this study, we demonstrate that the computational approaches are feasible to improve the Ang2-binding ability of aptamer. Although computational approaches can help reduce time and money consumption in the selection of aptamer, experimental screening procedures still need to verify the real target-binding ability of selected aptamers. In order to obtain better simulation results of protein-nucleic acid interactions, a distance dependent, knowledge-based coarse grained force field is recently developed for evaluating protein-DNA docking [61]. The force field can improve the quality of predictions in the protein-DNA docking, and they find shape complementarity and sequence-dependent DNA internal energy have great contribution to the specific protein-DNA interaction [61]. We think the synergy of computational techniques and experimental approaches can give great contributions in studying protein-nucleic acid interactions.

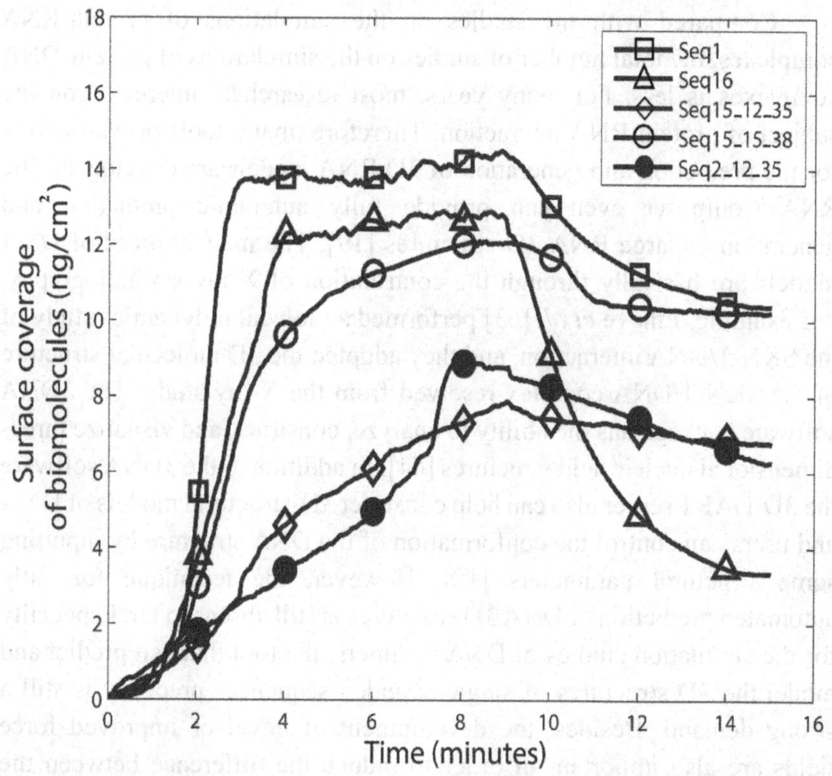

Figure 2. Curves of SPR experiments. Representative experimental curves measured on the SPR sensing platform for the interactions between Ang2 and different aptamers on sensor surface [62].

Table 1. Experimental data and the computationally obtained scores [62].

Name of aptamer	Surface coverage of biomolecules (ng/cm^2) (AVG±SD)	k_a ($\times 10^3$ M^{-1}s^{-1})	k_d ($\times 10^{-3}$ s^{-1})	K_A ($\times 10^6$ M^{-1})	*ZRANK score*
Seq1	11.17±1.47	10.02	1.39	7.23	–93.855
Seq16	1.87±0.31	1.66	4.99	0.33	–61.969
Seq15_12_35	8.12±0.61	6.03	0.61	9.89	–93.335
Seq15_15_38	11.69±1.11	8.22	0.97	8.47	–89.904
Seq2_12_35	5.68±0.41	4.07	0.79	5.15	-97.609

Compared with the studies on the simulations of protein-RNA complexes, the total number of studies on the simulations of protein-DNA complexes is less. For many years, most researchers interested on the subject of protein-RNA interaction. Therefore, many tools or webservers for the prediction and generation of 3D RNA models are developed. The RNA Composer even can provide fully automatic prediction and generation of large RNA 3D structures [19]. The main sources of DNA models are basically through the contribution of X-ray crystallography. For example, Etheve *et al.* [63] performed a molecular dynamics study of the SKN-1/DNA interaction, and they adopted the 3D molecular structure of the SKN-1/DNA complex resolved from the X-ray study. The 3DNA software package has the ability to analyze, construct, and visualize three-dimensional nucleic acid structures [64]. In addition to the 3DNA software, the 3D-DART server also can help construct 3D-structural models of DNA, and users can control the conformation of the DNA structure by inputting some structural parameters [18]. However, the technique for fully automated prediction of DNA 3D structures is still absent so far. Especially for the simulation studies of DNA aptamers, the tool that can predict and model the 3D structures of single-stranded sequences precisely is still a strong demand. Besides, the development of novel or improved force fields are also important in order to reduce the difference between the experimental and simulation results.

References

1. Alder, B. J., Wainwright, T. E. (1957). Phase transition for a hard sphere system, J. Chem. Phys., 27, pp. 1208–1209.
2. Alder, B. J., Wainwright, T. E. (1959). Studies in molecular dynamics. I. general method, J. Chem. Phys., 31, pp. 459–466.
3. Rahman, A. (1964). Correlations in the motion of atoms in liquid argon, Phys. Rev., 136, pp. A405–A411.
4. McCammon J. A. (1976). Molecular dynamics study of the bovine pancreatic trypsin inhibitor, In Models for Protein Dynamics, CECAM, pp. 137 (in France).
5. McCammon, J. A., Gelin, B. R., Karplus, M. (1977). Dynamics of folded proteins, Nature, 267, pp. 585–590.
6. Arnold, K., Bordoli, L., Kopp, J., Schwede, T. (2006). The SWISS-MODEL workspace: a web-based environment for protein structure homology modelling, Bioinformatics, 22, pp. 195–201.

7. Guex, N., Peitsch, M. C., Schwede, T. (2009). Automated comparative protein structure modeling with SWISS-MODEL and Swiss-PdbViewer: a historical perspective, Electrophoresis, 30 Suppl 1, pp. S162–S173.
8. Biasini, M., Bienert, S., Waterhouse, A., Arnold, K., Studer, G., Schmidt, T., Kiefer, F., Gallo Cassarino, T., Bertoni, M., Bordoli, L., Schwede, T. (2014). SWISS-MODEL: modelling protein tertiary and quaternary structure using evolutionary information, Nucleic Acids Res., 42, pp. W252–W258.
9. Kelley, L. A., Mezulis, S., Yates, C. M., Wass, M. N., Sternberg, M. J. E. (2015). The Phyre2 web portal for protein modeling, prediction and analysis, Nat. Protoc., 10, pp. 845–858.
10. Levitt, M. (1983) Computer simulation of DNA double-helix dynamics, Cold Spring Harb. Symp. Quant. Biol., 47 Pt 1, pp. 251–262.
11. Tidor, B., Irikura, K. K., Brooks, B. R., Karplus, M. (1983). Dynamics of DNA oligomers, J. Biomol. Struct. Dyn., 1, pp. 231–252.
12. Pérez, A., Marchán, I., Svozil, D., Sponer, J., Cheatham, T. E., Laughton, C. A., Orozco, M. (2007). Refinement of the AMBER force field for nucleic acids: improving the description of α/γ conformers, Biophys. J., 92, pp. 3817–3829.
13. Weiner, P. K., Kollman, P. A. (1981). AMBER: Assisted model building with energy refinement. A general program for modeling molecules and their interactions, J. Comput. Chem., 2, pp. 287–303.
14. Cornell, W. D., Cieplak, P., Bayly, C. I., Gould, I. R., Merz, K. M., Ferguson, D. M., Spellmeyer, D. C., Fox, T., Caldwell, J. W., Kollman, P. A. (1996). A second generation force field for the simulation of proteins, nucleic acids, and organic molecules J. Am. Chem. Soc. 1995, 117, 5179–5197, J. Am. Chem. Soc., 118, pp. 2309–2309.
15. Cheatham, T. E. 3rd, Cieplak, P., Kollman, P. A. (1999). A modified version of the Cornell *et al.* force field with improved sugar pucker phases and helical repeat, J. Biomol. Struct. Dyn., 16, pp. 845–862.
16. Tuszynska, I., Magnus, M., Jonak, K., Dawson, W., Bujnicki, J. M. (2015). NPDock: a web server for protein–nucleic acid docking. Nucleic Acids Res., 43(Web Server issue), pp. W425–W430.
17. Popenda, M., Błażewicz, M., Szachniuk, M., Adamiak, R. W. (2008). RNA FRABASE version 1.0: an engine with a database to search for the three-dimensional fragments within RNA structures, Nucleic Acids Res., 36, pp. D386–D391.
18. van Dijk, M., Bonvin, A. M. J. J. (2009). 3D-DART: a DNA structure modelling server, Nucleic Acids Res., 37(Web Server issue), pp. W235–W239.
19. Popenda, M., Szachniuk, M., Antczak, M., Purzycka, K. J., Lukasiak, P., Bartol, N., Blazewicz, J., Adamiak, R. W. (2012). Automated 3D structure composition for large RNAs. Nucleic Acids Res., 40, e112. doi: 10.1093/nar/gks339.
20. Bosshard, H. R., Marti, D. N., Jelesarov, I. (2004). Protein stabilization by salt bridges: concepts, experimental approaches and clarification of some misunderstandings. J. Mol. Recognit., 17, pp. 1–16.

21. Xu, D., Lin, S. L., Nussinov, R. (1997). Protein binding versus protein folding: the role of hydrophilic bridges in protein associations, J. Mol. Biol., 265, pp. 68–84.
22. Mandel-Gutfreund, Y., Schueler, O., Margalit, H. (1995). Comprehensive analysis of hydrogen bonds in regulatory protein DNA-complexes: in search of common principles, J. Mol. Biol., 253, pp. 370–382.
23. Jones, S., Shanahan, H. P., Berman, H. M., Thornton, J. M. (2003). Using electrostatic potentials to predict DNA-binding sites on DNA-binding proteins. Nucleic Acids Res., 31, pp. 7189–7198.
24. Chen, Y. C., Lim, C. (2008). Predicting RNA-binding sites from the protein structure based on electrostatics, evolution and geometry, Nucleic Acids Res., 36, e29. doi: 10.1093/nar/gkn008.
25. Dougherty, R. C. (1998). Temperature and pressure dependence of hydrogen bond strength: A perturbation molecular orbital approach, J. Chem. Phys., 109, pp. 7372–7378.
26. Greenwood, N. N. and Earnshaw, A. (1997). *Chemistry of the Elements*, 2nd Ed. (Elsevier Ltd, United Kingdom).
27. Sinden, R. R. (1994). *DNA Structure and Function.* Sinden, R. R., Chapter 1 "Introduction to the Structure, Properties, and Reactions of DNA," (Academic Press, San Diego) pp. 1–57.
28. Yakovchuk, P., Protozanova, E., Frank-Kamenetskii, M. D. (2006). Base-stacking and base-pairing contributions into thermal stability of the DNA double helix, Nucleic Acids Res., 34, pp. 564–574.
29. Haijun, Z., Zhong-can, O.-Y. (1999). Bending and base-stacking interactions in double-stranded semiflexible polymer, Phys. Rev. Lett., 82, pp. 4560–4563.
30. Theobald, D. L., Schultz, S. C. (2003). Nucleotide shuffling and ssDNA recognition in Oxytricha nova telomere end-binding protein complexes, EMBO J., 22, pp. 4314–4324.
31. Strick, T. R., Allemand, J. F., Bensimon, D., Bensimon, A., Croquette, V. (1996). The elasticity of a single supercoiled DNA molecule, Science, 271, pp. 1835–1837.
32. Saenger, W. (1984) *Principles of Nucleic Acid Structure*, 1st Ed. (Springer, USA).
33. Hornak, V., Abel, R., Okur, A., Strockbine, B., Roitberg, A., Simmerling, C. (2006). Comparison of multiple Amber force fields and development of improved protein backbone parameters, Proteins, 65, pp. 712–725.
34. Krepl, M., Cléry, A., Blatter, M., Allain, F. H. T., Sponer, J. (2016). Synergy between NMR measurements and MD simulations of protein/RNA complexes: application to the RRMs, the most common RNA recognition motifs, Nucleic Acids Res., 44, pp. 6452-6470. doi: 10.1093/nar/gkw438.
35. Brooks, B. R., Bruccoleri, R. E., Olafson, B. D., States, D. J., Swaminathan, S., Karplus, M. (1983). CHARMM: a program for macromolecular energy, minimization, and dynamics calculations, J. Comput. Chem., 4, pp. 187–217.
36. Bottaro, S., Lindorff-Larsen, K., Best, R. B. (2013). Variational optimization of an all-atom implicit solvent force field to match explicit solvent simulation data, J. Chem. Theory Comput., 9, pp. 5641–5652.

37. MacKerell, A. D., Bashford, D., Bellott, M., Dunbrack, R. L., Evanseck, J. D., Field, M. J., Fischer, S., Gao, J., Guo, H., Ha, S., Joseph-McCarthy, D., Kuchnir, L., Kuczera, K., Lau, F. T., Mattos, C., Michnick, S., Ngo, T., Nguyen, D. T., Prodhom, B., Reiher, W. E., Roux, B., Schlenkrich, M., Smith, J. C., Stote, R., Straub, J., Watanabe, M., Wiorkiewicz-Kuczera, J., Yin, D., Karplus, M. (1998). All-atom empirical potential for molecular modeling and dynamics studies of proteins, J. Phys. Chem. B, 102, pp. 3586–3616.
38. MacKerell, A. D. J., Banavali, N., Foloppe, N. (2000). Development and current status of the CHARMM force field for nucleic acids, Biopolymers, 56, pp. 257–265.
39. Klauda, J. B., Venable, R. M., Freites, J. A., O'Connor, J. W., Tobias, D. J., Mondragon-Ramirez, C., Vorobyov, I., MacKerell, A. D. J., Pastor, R. W. (2010). Update of the CHARMM all-atom additive force field for lipids: validation on six lipid types, J. Phys. Chem. B, 114, 7830–7843.
40. Raman, E. P., Guvench, O., MacKerell, A. D. J. (2010). CHARMM additive all-atom force field for glycosidic linkages in carbohydrates involving furanoses, J. Phys. Chem. B, 114, pp. 12981–12994.
41. Hatcher, E., Guvench, O., Mackerell, A. D. (2009). CHARMM additive all-atom force field for aldopentofuranoses, methyl-aldopentofuranosides, and fructofuranose, J. Phys. Chem. B, 113, pp. 12466–12476.
42. Vanommeslaeghe, K., Hatcher, E., Acharya, C., Kundu, S., Zhong, S., Shim, J., Darian, E., Guvench, O., Lopes, P., Vorobyov, I., MacKerell, A. D. (2010). CHARMM general force field (CGenFF): a force field for drug-like molecules compatible with the CHARMM all-atom additive biological force fields, J. Comput. Chem., 31, pp. 671–690.
43. MacKerell, A. D., Nilsson, L. (2008). Molecular dynamics simulations of nucleic acid-protein complexes, Curr. Opin. Struct. Biol., 18, pp. 194–199.
44. Schwede, T., Kopp, J., Guex, N., Peitsch, M. C. (2003). SWISS-MODEL: an automated protein homology-modeling server, Nucleic Acids Res., 31, pp. 3381–3385.
45. Pieper, U., Webb, B. M., Dong, G. Q., Schneidman-Duhovny, D., Fan, H., Kim, S. J., Khuri, N., Spill, Y. G., Weinkam, P., Hammel, M., Tainer, J. A., Nilges, M., Sali, A. (2014). ModBase, a database of annotated comparative protein structure models and associated resources, Nucleic Acids Res., 42, pp. D336–D346.
46. Bhattacharya, A., Wunderlich, Z., Monleon, D., Tejero, R., Montelione, G. T. (2008). Assessing model accuracy using the homology modeling automatically software. Proteins, 70, pp. 105–118.
47. Roy, A., Kucukural, A., Zhang, Y. (2010). I-TASSER: a unified platform for automated protein structure and function prediction, Nat. Protoc., 5, pp. 725–738.
48. Webb, B., Sali, A. (2016). Comparative protein structure modeling using MODELLER, Curr. Protoc. Bioinform., 54, pp. 5.6.1-5.6.37. doi: 10.1002/cpbi.3.
49. Lu, X.-. J., Olson, W. K. (2003). 3DNA: a software package for the analysis, rebuilding and visualization of three-dimensional nucleic acid structures, Nucleic Acids Res, 31, pp. 5108-5121.

50. Afzal, M., Shahid, A. A., Shehzadi, A., Nadeem, S., Husnain, T. (2012). RDNAnalyzer: A tool for DNA secondary structure prediction and sequence analysis, Bioinformation, 8, pp. 687–690.
51. Gruber, A. R., Lorenz, R., Bernhart, S. H., Neuböck, R., Hofacker, I. L. (2008). The Vienna RNA website. Nucleic Acids Res., 36, pp. W70–W74.
52. Reuter, J. S., Mathews, D. H. (2010). RNAstructure: software for RNA secondary structure prediction and analysis, BMC Bioinformatics, 11, 129, doi: 10.1186/1471-2105-11-129.
53. Popenda, M., Szachniuk, M., Blazewicz, M., Wasik, S.; Burke, E. K., Blazewicz, J., Adamiak, R. W. (2010). RNA FRABASE 2.0: an advanced web-accessible database with the capacity to search the three-dimensional fragments within RNA structures, BMC Bioinformatics, 11, 231. doi: 10.1186/1471-2105-11-231.
54. Bini, A., Mascini, M., Mascini, M., Turner, A. P. F. (2011). Selection of thrombin-binding aptamers by using computational approach for aptasensor application. Biosens. Bioelectron., 26, pp. 4411–4416.
55. Halgren, T. A. (1996). Merck molecular force field. I. Basis, form, scope, parameterization, and performance of MMFF94, J. Comput. Chem., 17, pp. 490–519.
56. Chen, R., Li, L., Weng, Z. (2003). ZDOCK: an initial-stage protein-docking algorithm, Proteins, 52, pp. 80–87.
57. Kumar, J. V., Chen, W. Y., Tsai, J. J. P., Hu, W. P. (2013). Molecular simulation methods for selecting thrombin-binding aptamers, Lect. Notes Electr. Eng., 253, pp. 977–983.
58. Hu, W. P., Kumar, J. V., Huang, C. J., Chen, W. Y. (2015). Computational selection of RNA aptamer against angiopoietin-2 and experimental evaluation, Biomed Res. Int., 2015, 658712. doi: 10.1155/2015/658712.
59. Sarraf-Yazdi, S., Mi, J., Moeller, B. J., Niu, X., White, R. R., Kontos, C. D., Sullenger, B. A., Dewhirst, M. W., Clary, B. M. (2008). Inhibition of in vivo tumor angiogenesis and growth via systemic delivery of an angiopoietin 2-specific RNA aptamer, J. Surg. Res., 146, pp. 16–23.
60. White, R. R., Shan, S., Rusconi, C. P., Shetty, G., Dewhirst, M. W., Kontos, C. D., Sullenger, B. A. (2003). Inhibition of rat corneal angiogenesis by a nuclease-resistant RNA aptamer specific for angiopoietin-2, Proc. Natl. Acad. Sci. U. S. A., 100, pp. 5028–5033.
61. Setny, P., Bahadur, R. P., Zacharias, M. (2012). Protein-DNA docking with a coarse-grained force field, BMC Bioinformatics 2012, 13, 228. doi: 10.1186/1471-2105-13-228.
62. Kumar, J. V., Tsai, J. J. P.; Hu, W. P., Chen, W. Y. (2015). Comparative molecular simulation method for ang2 / aptamers with in vitro studies, Int. J. Pharma Med. Biol. Sci., 4, pp. 61–64.

63. Etheve, L., Martin, J., Lavery, R. (2015). Dynamics and recognition within a protein-DNA complex: a molecular dynamics study of the SKN-1/DNA interaction, Nucleic Acids Res., 44, pp. 1440–1448.
64. Colasanti, A. V., Lu, X.-J., Olson, W. K. (2013). Analyzing and building nucleic acid structures with 3DNA, J. Vis. Exp., 2013, e4401. doi: 10.3791/4401.

Chapter 6

Bioinformatics analysis of microRNA and protein-protein interaction in plant host-pathogen interaction system

Nilubon Kurubanjerdjit

School of Information Technology, Mae Fah Luang University, Chiang Rai 57100, Thailand

Ka-Lok Ng

Department of Bioinformatics and Medical Engineering Asia University, Taichung 41354, Taiwan

Department of Medical Research China Medical University Hospital China Medical University, Taichung 40402, Taiwan

6.1 Introduction

Plant systems are continuously subjected to attack by pathogens, which resulted in severe damage and huge economic cost in agriculture. Therefore, it is important to elucidate the host-pathogen interaction in plant systems. *Arabidopsis thaliana*, a long day plant, is a well-known model organism for plant science [Mandoli D.F. 2000]. *A. thaliana* is chosen for most of the studies because of two main reasons: i) The whole genome sequence was determined since 2000; and ii) There are many molecular biology lab techniques, such as cDNA, genomic libraries, bacterial artificial chromosomes, microarrays and ESTs, are available for the study of its biological processes, mechanisms and functions [Mandoli D.F. 2000].

MiRNAs (miRNA) are a class of small non-coding RNAs, found in eukaryotic cells (Bartel 2004), that bind to mRNA and induce either translation repression or mRNA degradation. The translational inhibition by miRNAs has been thought of as a major mechanism in animal systems while mRNA degradation or post-transcriptional regulation has been considered as a major regulatory mechanism in plants [Dai X 2010]. MiRNAs play crucial roles in *A. thaliana* biological processes, such as leaf sidedness, flower development, hormone signaling and metabolism, and both biotic and abiotic stress response. Previous studies had tried to clarify the functions of miRNA using experimental and computational approaches. In target prediction approaches, multiple factors are introduced to identify miRNA target genes i.e. complementarity of different regions on miRNA, binding site conservation and target site hybridization free energy and accessibility. Generally, a pathogenic bacterium attacks hosts in many ways including sticking and colonizing host tissues, secreting degradation enzymes and toxins release. Many of such mechanisms involve host-pathogen protein-protein interaction (PPI). PPI is an essential process of living cells [Lin N. 2004]. It also plays a crucial role in some critical interspecies interactions such as host-pathogen interactions and pathogenicity [Casadevall A. 2000]. Recently high throughput proteomic technology has uncovered a large number of PPI, particularly in interspecies protein interactions of plants and bacteria [Tsao T.H. 2011]. Therefore, comprehensive knowledge of host-pathogen PPI and interactome analysis can help accelerating protein annotations and elucidate a plant's immune system against bacteria.

Recently, PPI networks in model organisms such as *Saccharomyces cerevisiae* [Ito T. 2000; Uetz P. 2000; Ito T. 2001] and *Escherchia coli* [Arifuzzaman M. 2006] have been reported. Protein interactions appear to form a molecular network, which usually contains small circuit patterns called network motifs. The proteins of a network motif are usually involved in similar biological processes, and protein complexes can be identified by clustering the network [Palla G. 2005; Jonsson P. 2006].

However, few have been known for plant-pathogen interactions. Flor [Flor HH. 1971] proposed a theory on the PPI between pathogen effector protein accessibility, but it is still unclear which factors are significant or have much influence on target prediction. Thus, most of the current predictive approaches are based on certain factors, which may affect the accuracy of prediction. Therefore, integrating diverse tools may potentially improve target prediction.

Xanthomonas campestris pv campestris (*Xcc*) is one of the pathogenic gram-negative bacteria that cause blights and rots in plants [Tsuji J. 1988; Tsuji J. 1991; Tsuji J. 1992; Buell C.R. 2002]. Host infections caused by *Xcc* can occur in any stage of the plant life cycle. Symptoms resulted from this pathogen have been reported in many previous research works.

6.2 The conceptual framework of this study

Figure 1 depicts overview system of this research work. A unified dataset comprised of *Xcc* effectors, *A. thaliana* miRNAs, target genes and PPI was employed for the host-pathogen interaction study. The study is divided into three parts as following:

I. Prediction of miRNA-regulated protein PPI pathways in *A. thaliana* using machine learning algorithms [Kurubanjerdjit N. 2013a]

II. Prediction of PPI for *A. thaliana* and *Xcc* based on protein domain-domain interaction (DDI) and interolog approaches [Kurubanjerdjit N. 2013b]

III. Investigate the binding interaction between *Xcc* effectors and *A. thaliana* miRNA promoters [Kurubanjerdjit N. 2014]

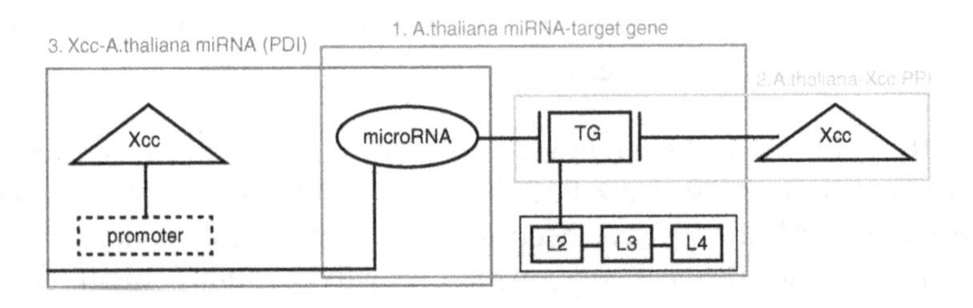

Fig. 1 Conceptual framework of the study [Kurubanjerdjit N. 2014]

6.2.1 Study I: Prediction of miRNA-regulated PPI pathways in A. thaliana using machine learning algorithms

We identify *A. thaliana* miRNAs target prediction by integrating scores from three miRNA target prediction tools: PITA, miRanda and RNAHybrid algorithms, and then made use of three machine learning algorithms, i.e. Support Vector Machine (SVM), Random Forest tree (RF) and neural network (NN), to infer final predictions by majority voting. Then, miRNA-regulated PPI pathways are constructed by linking the plant resistance genes (PRG) and transcription factor (TF) from the PRGdb [Sanseverino W. 2010] and the TRANSFAC® database database [Matys V 2006] respectively. These miRNA-regulated pathways would provide new insights into the plant pathogen interaction networks. Downstream pathways are characterized by the Jaccard coefficient, which is implemented based on Gene Ontology.

6.2.1.1 *Data Sources*

There are three resources that we used in this study i) a dataset of 563 confirmed *A.thaliana* miRNA-target pairs was obtained from the *Arabidopsis* Small RNA project Database (ASRP) [Gustafson A.M. 2005] which comprise the interactions of 118 miRNAs and 205 mRNAs ii) a set of 243 *A. thaliana* miRNAs with their sequences obtained from miRBase [Griffiths-Jones S. 2008] and iii) a gene set (33,539 genes) with their mRNA FASTA sequences collected from The *Arabidopsis* Information Resource (TAIR version 10) [Rhee SY 2003].
Besides, the genomic 3'UTR information of *A.thaliana* was extracted from TAIR. Furthermore, Dinucleotide statistical information of 3'UTR was obtained from Genomatix Software GmbH Company located at Munich (see http://www.genomatix.de/). Those two pieces of information are for RNAHybrid calculation. Moreover, the gene annotation information was carried from the GO website.

6.2.1.2 *Dataset Preparation*

For training dataset, experimentally confirmed miRNA-target pairs from ASRP (BLAST e-values are somewhere between $2*10^{-10}$ and 0.62) was processed by the three predictors that used the three algorithms' default

parameter setting. The positive training set (406 pairs) are experimentally confirmed pairs. The negative set, a total of 9938, comprised pairs that satisfied the three algorithms' default parameter settings with the positive set subtracted.
Moreover, the test set was generated by combining the three prediction scores with their default parameter setting for a set of 243 *A. thaliana* miRNA and a genome wide set of UTR.

6.2.1.3 *System Workflow*

A system was set up to predict miRNA target genes of *A. thaliana.* Finally, the predicted targets were linked with their PRG and TFs to establish the pathway database. Figure 2 depicts system workflow of the study.

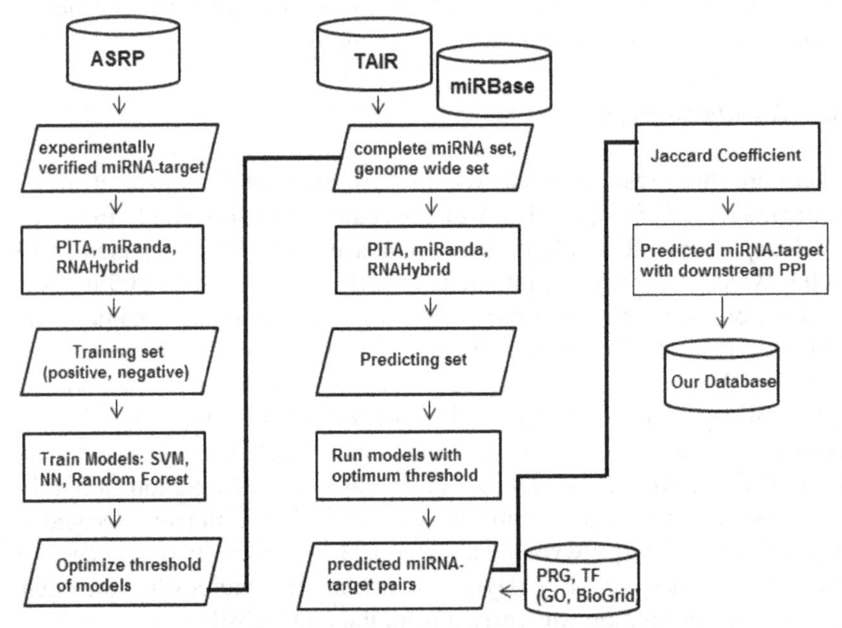

Fig. 2 miRNA-target Prediction System Flowchart [Kurubanjerdjit N. 2013a]

6.2.1.4 *Prediction of miRNA target genes*

The prediction score of RNAHybrid is derived from the mean free energy (MFE) value of binding between the miRNA and the target gene.

In case RNAHybrid returns multiple MFE values with the same miRNA-target gene interaction, the target score is given by Equation 1,

$$TARGET\_SCORE = -\log(\sum(e^{-mfe})) \quad (1).$$

For PITA, default parameter settings were used. The ΔΔG value is used in case of single binding sites occurrences, while the determined prediction score by Eq. 1 is used in case of multiple binding sites, where *mfe* is replaced by ΔΔG, which denotes the free energy value of binding between the miRNA and the target gene.
For MiRanda, *max score* is considered. There are three parameters are required; the threshold score, MFE, and scaling factor are set to 80, -14 kcal/mol and 2.0 respectively.

6.2.1.5 *Use of machine learning algorithms for miRNA target classification*

In order to achieve the best performance, optimal parameter settings were identified [Kurubanjerdjit N. 2013a] for the three classifiers, i.e. SVM, RF and NN. Once the classifier was optimized, it was adopted to evaluate the prediction.
Then, the three classifier prediction results were integrated and categorized into four groups; (i) 3-vote which is composed of the miRNA-target binding pair prediction result from the three classifiers, (ii) 2-vote in which either of two classifiers predict as a binding pair, (iii) 1-vote in which only one classifier predicted as a binding pair, and (iv) 0-vote where the test pair does not belong to any classifier.

6.2.1.6 *MiRNA-regulated PPI pathways*

To quantify the relationship among miRNAs, target genes, and their PPIs, the importance of miRNA-PPI coupled networks are ranked by performing enrichment analysis. Enrichment analysis was performed by computing the Jaccard coefficient (*JC*) to rank the significance of such relations. Let a miRNA target pathway be denoted by miRNA–TG–L2, where TG denotes target gene or level one protein (L1) GO annotation, and L2 denotes level two proteins (L2) annotation. The JC for the pathway TG–L2, is given by JC(TG, L2).

6.2.1.7 *Results*

Optimizing and performance comparison of target prediction tools

Our finding indicates that the optimal threshold prediction score of each predictor; ΔΔG is -20 kcal/mol (PITA), *max score* is 116 (miRanda), and MFE is -28 kcal/mol (RNAHybrid). In term of the prediction performance, RNAHybrid gave the best sensitivity (0.938) and *F1*-score (0.927), whereas PITA (0.950) gave the greatest specificity.

In order to observe the prediction performance of each individual predictor and the combination approach, ROC known as the AUC [Kim S.K. 2006] and the five performance measures (accuracy, sensitivity, specificity, *F1*-score and Matthew correlation coefficient) were examined. Since our training dataset is unbalance set which are 406 and 9938 miRNA-target binding pairs of positive and negative set respectively. AUC values are reported for both the original imbalance and also balanced cases (406 and about 1,000 miRNA-target binding pairs of the positive and negative dataset). To generate the balanced case, the 406 positive target interactions were kept, and a total of 1000 entries, around 2.5 times large as the positive set, were randomly selected from the negative set. The AUC values of the three predictors' features and the combination approach show that the use of balanced set of features combination achieved better AUC value, although inclusion of poor feature could slightly degrade the performance [Kurubanjerdjit N. 2013a].

Furthermore, the five performance measures are calculated to compare the prediction performance of the three predictors and their combinations. Our finding indicates that the use of three predictors' features gave the best performance in most of the cases. In fact, among the five measures, combination of the three predictors' scores attained the best performance for two of them, i.e. ACC and MCC measures. In addition to the AUC study, this results support our hypothesis that the use of multiple features may improve the overall performance.

Performance comparison of classifiers and combinations of classifiers

Since the combination of the three predictors' scores may improve the overall performance, we assume this in the following study.

There may be concerned that combination of classifiers may resulted in increasing the number of FP events. Ten-fold cross-validation tests were

conducted. It was found the combination of NN, SVM and RF classifier gave the second lowest FP events [Kurubanjerdjit N. 2013a].
Moreover, a 10 fold cross-validation test was performed in order to test the results robustness in which the 1,000 records of the negative training set were randomly selected; thus, the standard deviation for the five performance measures were computed. Our finding indicates that there is a small fraction of the standard error (0.0124) in every measure; this evidence indicates that the performance is rather stable with respect to randomization [Kurubanjerdjit N. 2013b].
MicroRNA target gene interaction predictions
Genome-wide miRNA target prediction was performed using the three target predictors, where feature vectors of the 158,750 miRNA-target interactions were input into the NN, SVM and RF classifiers. As a result of classification, 3-vote, 2-vote and 1-vote groups were identified [Kurubanjerdjit N. 2013b].

6.2.2 Study II: Prediction of PPI between *A. thaliana* and *Xcc* based on protein DDI and interolog approaches

Interactions of *A. thaliana* proteins and *Xcc* pathogen bacteria proteins are identified by two different approaches; the DDI approach which infers interspecies PPIs by known DDI recorded by various databases; and the interolog approach that identifies PPIs based on homologous pairs of PPI across different organisms. The results from these two methods are integrated and the dense protein interaction regions are specified by clique percolation network analysis. In particular, the PRG information and the bacterial effector proteins are studied to provide new insights into the molecular mechanism of the host-pathogen system.

6.2.2.1 *Data Sources*

To perform the DDI approach, A list of 46,991 *A.thaliana*'s PFam domains was extracted from TAIR (http://www.arabidopsis.org) version 10 with filtering out a set of domain that satisfy the 10^{-4} e-value BLAST, and the aligned sequence length coverage is set to 90%. Besides, a list of 304 Xcc domains was extracted from UniProt, (http://www.uniprot.org). Furthermore, the DDI information was obtained from the two difference resources which are iPFam (http://ipfam.sanger.ac.uk) and 3DID (http://3did.irbbarcelona.org).

For interolog approach, a set of 63,830 experimentally confirmed PPI was gathered from DIP (http://dip.doe-mbi.ucla.edu/dip/main.cgi). A list of 35,386 *A.thaliana* protein sequences was gathered from TAIR, while a list of 202 *Xcc* protein sequences was obtained from Uniprot. For building up the community of interacting *A.thaliana* PPI, a set of 7,466 experimentally confirmed *A.thaliana* interactomes was obtained from two difference sources which are TAIR and BioGrid (http://www.thebiogrid.org). Furthermore, PRG and TF were retrieved from the PRGdb and TRANSFAC® sources respectively.

6.2.2.2 *System Workflow*

Host-pathogen PPI was identified using two pipelines; DDI and the interolog approaches as shown in the system flowchart as Fig. 3. Subsequently, the set of predicted PPI from both methods was subjected to enrichment analysis via DAVID [Huang D.W. 2009]. In addition, information about PRG, TF and *Xcc* effector proteins was integrated into our system.

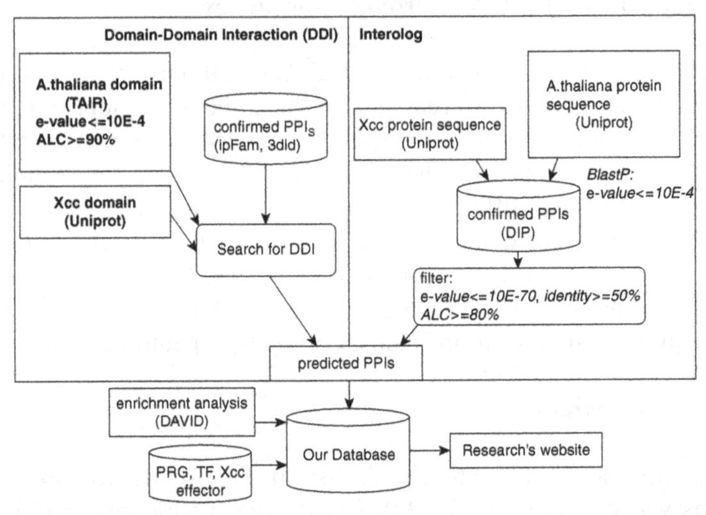

Fig. 3 System Flowchart of PPI prediction of *A. thaliana* and *Xcc*. Prediction is based on the (i) domain-based and, (ii) interolog approaches [Kurubanjerdjit N. 2013b]

6.2.2.3 *PPI prediction by DDI*

The interacting protein pair of *A.thaliana* and *Xcc* could be identified based on the assumption that the interspecies PPI contains an interacting

PFam domain pairs. The known DDI were gathered by the iPFam and 3DID databases. The three sets of input were merged: i) a set of 1,555 *A. thaliana* PFam domains, ii) a total of 304 *Xcc* PFam domains, and iii) a collection of 7,039 known DDI recorded by iPFam and 3DID.

6.2.2.4 *PPI prediction by the interolog approach*

An interolog is a conserved interaction between a pair of proteins which have interacting homologs in another organism. *A.thaliana* protein sequences from TAIR and Xcc protein sequences from uniprot were blasted against a set of known PPI from DIP database with the e-value cutoff, sequence identity bound and aligned sequence length coverage set to 1.0×10^{-70}, 50% and 80% respectively. The cutoff values was carried forward from the work of Yu [Yu H. 2004] and our cutoff values is more strict than the work of Li [Li Z.G. 2011] which predicted the interspecies PPI of bacterium *R. solanacearum* and *A. thaliana* by the interolog approach. Finally, the host-pathogen PPI were identified if a protein pair of *A. thaliana* and *Xcc* has the corresponding homologs in the DIP database.

6.2.2.5 *Xcc effector protein prediction*

The *Xcc* effector protein and the type III secretion system effector protein were identified in this study. To identify effector protein, a set of *Xcc* proteins was submitted to EffectiveT3 (http://www.effectors.org/) [Jehl M.A. 2011] with the default parameter setting: i) the organism type was set to gram-negative; ii) the classification module was set to type III effector prediction of the plant set; and iii) the cutoff default setting was set to 0.999; and iv) the domain score was set to 4.0.
Furthermore, to identify the type III secretion system effector protein, a set of Xcc proteins was submitted to ModLab system for identifying the existence of type III secretion system (T3SS) signals in amino acid sequences. The default parameter setting is used in the prediction: i) the prediction method was set to "neural network;" ii) the sequence truncation: N and C terminals were set to 1 and 30 respectively; and iii) the neural network threshold was set to 0.4.
Besides, the type III secretion system effector predicted by ModLab (Molecular Design Laboratory: http://gecco.org.chemie.uni-frankfurt.de/index.html) [Lower M. 2009] is a prediction system for identifying the existence of type III secretion system (T3SS) signals in

amino acid sequences. A set of *Xcc* protein sequences was input into the system, where the parameters were set to default values: i) the prediction method was set to "neural network;" ii) the sequence truncation: N and C terminals were set to 1 and 30 respectively; and iii) the neural network threshold was set to 0.4.

6.2.2.6 *Clique analysis of the PPI network*

The nodes of a cluster are usually involved in similar biological processes, and protein complexes can be identified through the clustering of a PPI network [Palla G. 2005; Jonsson P. 2006]. In order to investigate the functional modules in which potential pathogen-targeted *A. thaliana* proteins are involved, A set of experimentally confirmed *A.thaliana* interactomes was analyzed based on the clique percolation clustering approach by CFinder software [Adamcsek B. 2006]. The 3-community ($k=3$) was preliminary considered to analyze a PPI topological network. The 3-community, which contains at least one predicted *A. thaliana* protein that interacts with *Xcc,* were filtered and kept for enrichment analysis by using DAVID with the e-value cutoff set to 0.05.

6.2.2.7 *Results*

The A. thaliana and Xcc PPI networks

From our finding, a total of 1,011 possible host-pathogen PPIs was identified which composes of 398 *A. thaliana* proteins and 57 *Xcc* proteins. There are 241PPIs were predicted by domain-based approach and 913 PPIs were derived from the interolog approach. Moreover, Ten PPIs were consistently derived from both methods. Of the consistently predicted PPI, five PPI predicted by the DDI approach were confirmed by both the iPFam and 3DID. The number of consistent predicted PPI, i.e. 10, found in our work is close to that was predicted in the work of Li [Li Z.G. 2011], i.e. 12, which used the interolog and DDI methods to predict the PPI between *R. solanacearum* and *A. thaliana* proteins.
Our result indicates that an *Xcc* protein has about 18 *A. thaliana* interacting partners on average, while an *A. thaliana* protein interacts with around 2.5 *Xcc* proteins. This evidence is consistent with a scenario in which a few pathogenic proteins attack the host's proteome [He F. 2008; Li Z.G. 2011]. In addition, the work of Stahl and Bishop [Stahl

E.A. 2000] indicates that a pathogen mutates its genes to infect the host. However, the plant defends the attacks by expanding its gene families.

Xcc effector predicted by EffectiveT3 and the type III secretion system effector prediction (ModLab) system

As a result of implementing the EffectiveT3 tool, there are two *Xcc* proteins (P58892 and Q8PC32) are predicted as bacterial secreted proteins. Q8PC32 (dsbB) had been reported in the work of Jiang [Jiang B.L. 2008] that a mutation in the *dsbB* gene can result in ineffective type II and type III secretion systems. In addition, another two *Xcc* proteins (Q8PBK7 and P22260) are predicted as type III effector proteins. The work of Hsiao [Hsiao Y.M. 2005] demonstrated that P22260 (Clp) up-regulates the transcription of the engA gene encoding a virulence factor in *Xcc* by a direct binding to the upstream tandem Clp sites.
The ModLab software identified type III effector proteins (Q8P7S1, P22260, Q8PAK9 and Q8P815). Interestingly, P22260 (CRP-like protein) was identified as a type III effector protein by both predictors and also it was recorded by UniProt as a pathogenesis effector protein, as it undergoes specific processes that generate the ability of an organism to cause disease. Furthermore Q8P815 involves in the plant-pathogen interaction pathway recorded by KEGG. This finding also consists with the report of Buell [Buell C.R. 2002] indicating that the genes involved in the resistance response can be classified into three classes: (1) R genes which are involved in the recognition of the pathogen; (2) signal transduction genes; and (3) defense response genes which are involved in the suppression of pathogen development.

The results of Gene Ontology enrichment analysis

Our finding indicates the top three over-represented biological processes of *Xcc* proteins are enriched in i) the generation of precursor metabolites and energy; (ii) the protein folding process; and iii) the carboxylic acid biosynthetic process. While the *A. thaliana* proteins involved in the predicted PPI are mainly enriched in i) oxidation and reduction; ii) protein folding; and iii) response to cadmium ion which is consistent with the report from the work of Fones et al [Fones H. 2010] stating that responding to cadmium ion is a significant strategy adopted by plant defense against pathogen.
Our result also demonstrated that *Xcc* proteins tend to interact with a group of *A. thaliana* proteins involved in the same biological process.

For instance, i) P63447 interacts with 60 *A. thaliana* proteins all are involved in the oxidation reduction process; and ii) the type III effector Q8PAK9 interacts with 21 proteins of *A. thaliana* proteins all are involved in the process of responding to cadmium ion.

The results of clique network analysis

From our result, a total of 116 3-communities were obtained, of which eight communities contain *A. thaliana* proteins targeted by *Xcc* and these communities are referred to as pathogen-targeted communities in this study. The over-represented biological processes of *A. thaliana* proteins in these eight communities whose satisfied p-value cutoff was set to 0.05 are: i) those involving photosynthesis, ii) those responding to steroid hormone stimulus, steroid hormone mediated signaling, inorganic substances, metal ion, regulation of the cell cycle process, brassinosteroid mediated signaling, and the generation of the precursor metabolites and energy.
Besides, a 9-community, which is the highest *k* degree, was identified by Cfinder. Our evidence shows that the 9-community is composed of four *Xcc* effectors (P22260, Q8PD23, Q8P494, Q8PAV3) and six *Xcc* interacting partners (AT3G54180, AT1G66750, AT1G18040, AT4G28980, AT1G76540, AT3G48750).

6.2.3 Study III: Investigate the binding interaction between *Xcc* effectors and *A. thaliana* miRNA promoters

The inter-species interaction between an *Xcc* pathogen effector and *A. thaliana* miRNA transcription promoter was investigated using three methods: (i) interolog, (ii) alignment based on using transcription factor binding site (TFBS) profile matrix, and (iii) the web-based binding site prediction tool, PATSER. Furthermore, the results obtained from 'Study I' and 'Study II' were integrated into 'Study III'.

6.2.3.1 *Data sources*

To perform the interolog approach, a list of 202 Xcc protein sequence was extracted from UniProt, a list of 187 *A.thaliana* miRNA promotor sequences (3 kb upstream of the miRNA transcription region) was downloaded from the Plant MiRNA Database (PMRD) [Zhang Z. 2010]. Moreover, the experimentally confirmed PDI were extracted from

Protein-DNA Interface Database (PDIdb: http://melolab.org/pdidb/web/content/home) [Norambuena T. 2010].

To perform PDI prediction of *A.thaliana* and *Xcc*, aline the miRNA promotor sequence against the TFBS profile matrix was adopted. A list of 164 TFBS profile matrices from a variety of prokaryotic TF were obtained from PePPER [de-Jong H. 2012], and their protein sequences were obtained from UniProt.

6.2.3.2 *System workflow*

The interaction between *Xcc* effector and *A. thaliana* miRNA were identified using two pipelines as shown in system flowchart in Figure 4. Firstly, potential PDIs were predicted using the interolog approach and then the predicted PDI of *Xcc* effector and *A. thaliana* miRNA promoter was identified using the TFBS profile matrix of *Xcc* homolog proteins. In addition, three different types of interactions were identified and integrated to examine the interaction between *A. thaliana* and *Xcc*, which are (i) binding of *Xcc* effectors with *A. thaliana* miRNA promoter regions, (ii) the interaction of *A. thaliana* miRNA and their target gene, and (iii) the PPI of *A. thaliana* and *Xcc* proteins.

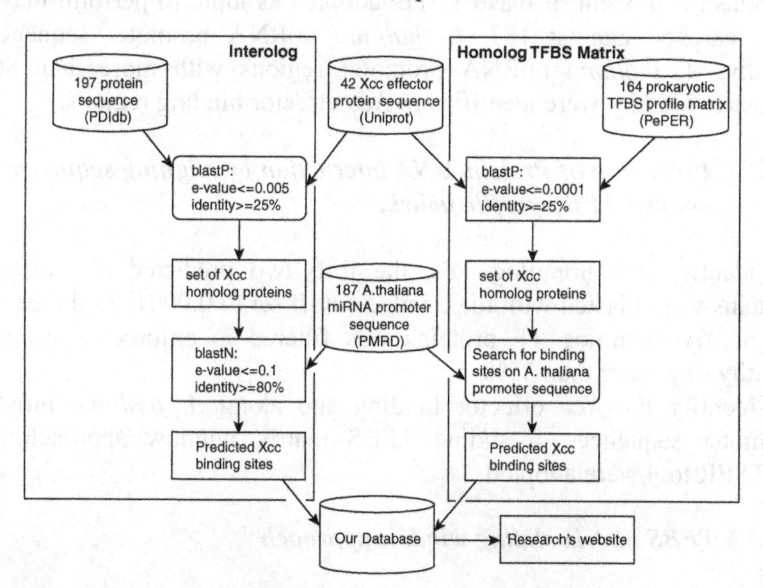

Fig. 4 System flowchart of PDI prediction between *Xcc* effector and *A. thaliana* miRNA. Prediction was based on the (i) interolog and, (ii) alignment of homolog TFBS profile matrix approaches [Kurubanjerdjit N. 2014].

6.2.3.3 *Prediction of protein-DNA interaction by interolog approach*

To identify the *Xcc* effector protein, EffectiveT3 and ModLab were employed.
BLAST was adopted to search for *Xcc* effector binding sites on *A. thaliana* promoter regions. Short alignments have relatively high e-values, this is because the calculation of the e-value takes into account the length of the query sequence. In general, one sets the e-value to be 0.0001 for blastP and blastN search [Claverie J.M. 2006]. However the protein sequences obtained from PDIdb have an average length of 116 bases long, whereas an average length for homologous TFBS study have an average of 404 bases long, therefore we set the e-value to be 0.005 for blastP search. And the e-value of 0.1 is set for the blastN search, because the average binding DNA sequence length from PDIdb is 15 bases long.
To perform blastP, forty-two predicted effector proteins were blasted against 197 proteins of various species obtained from PDIdb. A filter was applied to this set of *Xcc* homolog proteins to ensure greater than 25% sequence identity.
Subsequently, the interacting nucleotide sequences of the *Xcc* homolog proteins (as a result of blastP) were adopted as input to perform blastN, and searched against 187 *A. thaliana* miRNA promoter sequences. Finally, *A. thaliana* miRNA promoter regions with more than 80% sequence identity were identified as *Xcc* effector binding regions.

6.2.3.4 *Prediction of Protein-DNA interaction by aligning sequence against TFBS profile matrix*

To identify *Xcc* homolog TFs, the forty-two predicted *Xcc* effector proteins were blasted with the e-value cutoff set to 0.0001. And then this set of *Xcc* homolog TF proteins was filtered to enforce a sequence identity of greater than 25%.
To identify the *Xcc* effector binding site along *A. thaliana* miRNA promoter sequences, a sliding TFBS matrix window approach and PATSER tool were adopted.

6.2.3.5 *TFBS matrix sliding window approach*

The structure of TFBS matrix is a composite of four rows of the A, C, G, T and the column represents the weight occurrence scores in each position. The TFBS matrix window is applied by sliding it across aligned promoter sequences 1 bp at a time in order to sum up the score of hits that are presented in the window of all position. For each miRNA promoter sequence, the region which gives the highest matching score is assigned as the best binding region. Finally, only the binding regions which satisfied the 80% identity bound were selected. The identity measure is given by α/β, where α denotes the matching score of the binding region and β denotes the maximum matching score of the TFBS profile matrix.

6.2.3.6 *PATSER Tool*

In order to analyze the matching positions on miRNA promoter sequence according to TFBS matrix profiles, RSA-Tools-PATSER which is tool for identifying putative matching positions by scanning a DNA sequence with a position-specific scoring matrix (PSSM) was adopted. A set of 42 predicted *Xcc* effector proteins and also the TFBS profile matrices were submitted to PATSER (http://rsat.ulb.ac.be/patser_form.cgi) with default parameter setting. Finally, a set of binding sites in which the score was greater than seven was obtained.

6.2.3.7 *Prediction of TFBS at the miRNA host gene promoter region*

As an assumption that if TFBS has been found along miRNA promoter sequences, it is expected that such a region may be the binding site of a pathogen effector. In order to identify TFBS on promoter regions of miRNA host genes, the information of *A. thaliana* TFBS motif which is gathered from two resources; *A. thaliana* Promoter Binding element Database (AtProbe) and *Arabidopsis* Gene Regulatory Information Server (AGRIS). Subsequently, this information was searched along miRNA promoter sequence in order to find the matching regions.

6.2.3.8 *Gene Ontology Enrichment Analysis*

In order to identify the functional annotation of miRNA target genes where the promoter region is predicted to have PDI with the *Xcc* effector protein, a list of miRNA target gene which is potentially regulated by the

Xcc effectors was submitted to DAVID for clustering of the redundant annotation terms. Thus, enriched biological processes related gene lists were obtained.

6.2.3.9 ***Results***

The results of protein-DNA interaction predictions

Our finding identifies a list of 3,988 putative effector binding sites on *A. thaliana* miRNA promoters which involved three effectors and 187 miRNAs. The number of binding sites predicted by the interolog approach, alignment of homolog TFBS profile matrix by sliding window and by PATSER are 9, 3368 and 611 repectively. Furthermore, 42 *Xcc* effector proteins were identified in this study. Among the predicted *Xcc* effector proteins, P22260 (*clp*), which was predicted to bind to many miRNA promoter regions, is the type III effector protein involved in the pathogenesis process. This evidence is also consistent with the record in Uniprot, which indicates that this effector can enhance the ability of an organism to cause disease in another.

The results of enriched biological processes of miRNA target genes

The over-represented biological processes of miRNA target genes where the miRNA is predicted to have PDI with *Xcc* effector are mainly enriched in (i) response to cadmium ions, (ii) response to metal ions, and (iii) in response to inorganic substances.

Xcc effector binding sites at the A. thaliana miRNA promoter regions

From our results there are multiple regions along the *A. thaliana* miRNA promoters that were targeted by two *Xcc* effectors, i.e. P22260 and Q8P8F9.Three of the miRNAs, ath-MIR-408, ath-MIR-169i, and ath-MIR-169j are upstream regulators of two PRG [Kurubanjerdjit N. 2014].

The results of cis-regulatory binding site predictions

The locations of cis-regulatory binding sites determine the connectivity of genetic regulatory networks. Identification of these binding sites would facilitate research in certain key areas, including evolution, development and the pathogenic immunity system. Cis-regulatory binding sites on miRNA promoters where a number of TF can bind were observed in our previous study [Kurubanjerdjit N. 2014].

MiRNA targets predicted by the sliding window approach and PATSER

Our results indicated that a set of miRNAs, ath-MIR164c, ath-MIR167a, ath-MIR167b, ath-MIR169i, ath-MIR169j, ath-MIR169k, ath-MIR169l, ath-MIR408, ath-MIR771, ath-MIR843 and ath-MIR854a, target the *PRG*, *RIN1*. This gene has been reported by TAIR to function in regulation of defense responses to fungi, incompatible interactions, and regulation of flower development [Rhee SY 2003]. The results of the binding sites sequence information predicted by both the sliding window approach and PATSER for effectors P22260 and Q8P8F9, had been reported in a previous work by Kurubanjerdjit [Kurubanjerdjit N. 2014].

6.3 Discussion and Conclusion

In this work, a novel *A. thaliana* miRNA target prediction strategy was demonstrated based on integrating prediction scores from three target predictors (PITA, miRanda, RNAHybrid) to form a feature vector for three classifiers to conduct prediction by majority voting. It has been demonstrated that the combination those three classifiers achieved the best performance according to the ACC and MCC measures.
Furthermore, the interspecies PPI of *A. thaliana* and *Xcc* was identified based on the DDI and interolog approaches. The PPI network motifs were specified by the concept of clique percolation, and the biological processes of both *A. thaliana* and *Xcc* that are involved in the network motifs were observed. The PRG and also the effector bacterial protein were focused in our study. Our results suggested that a pathogen employs five strategies to achieve this goal. First, a few *Xcc* proteins tend to interact with *A. thaliana's* hub proteins or the PRG. Second, some of the *Xcc* proteins tend to interact with many *A. thaliana* proteins and third, it was found that some *Xcc* proteins target at a group of *A. thaliana* proteins involved in the response to cadmium ions, which is a significant plant biological process against pathogen. Fourth, many *Xcc* proteins target a few *A. thaliana* proteins involved in the plant-pathogen interaction pathways such as the response to the cadmium ion pathway and the defense response to bacterium, fungus and virus pathway. Fifth, the pathogen may make use of a specific kind of protein, i.e., a type III effector protein, to reprogram the host PPI.

In addition, the interspecies PPI of *A. thaliana* and *Xcc* was identified based on the DDI and interolog approaches. The PPI network motifs were specified by the concept of clique percolation, and the biological processes of both *A. thaliana* and *Xcc* that are involved in the network motifs were observed. A web site has been set up which is freely accessible at: http://ppi.bioinfo.asia.edu.tw/EDMRP. The importance of this web site can be understood in terms of the following information, (i) predicted interaction between pathogen effector and the host's miRNA promoter binding sites, (ii) putative regulation relationship between miRNA, PRG and TF, as well as the miRNA-regulated PPI pathways, and (iii) the host-pathogen PPI information.

References

Adamcsek B., Pella. G., Farkas IJ., Derenyi I., Vicsek T., (2006). "CFinder: locating cliques and overlapping modules in biological networks." *BMC Bioinformatics* 22: 1021-1023.

Arifuzzaman M., Maeda M., Itoh A., Nishikata K., Takita C., Saito R., Ara T., Nakahigashi K., Huang H. C., Hirai A., Tsuzuki K., Nakamura S., Altaf-Ul-Amin M., Oshima T., Baba T., Yamamoto N., Kawamura T., Ioka-Nakamichi T., Kitagawa M., Tomita M., Kanaya S., Wada C., Mori H., (2006). "Large-scale identification of protein-protein interaction of escherichia coli k-12." *Genome Research* 16(5): 686-691.

Babitha M.P., Bhat. S. G., Prakash H.S., Shetty H.S., (2002). "Differential induction of superoxide dismutase in downy mildew resistant and susceptible genotypes of pearly millet." *Plant Pathol* 15(4): 480-486.

Banjerdkit P., Vattanaviboon P., Mongkolsuk S., (2005). "Exposure to cadmium elevates expression of genes in the OxyR and OhrR regulons and induces cross-resistance to peroxide killing treatment in Xanthomonas camertis." *Appl Environ Microbiol* 71(4): 1843-1849.

Bartel, D. P. (2004). "MicroRNAs: genomics, biogenesis, mechanism, and function." *Cell* 116(2): 281-297.

Buell C.R. (2002). "Interactions between xanthomonas species and *Arabidopsis thaliana*." *Arabidopsis Book* 1: e0031

Casadevall A., Pirofski, L. A. (2000). "Host-pathogen interactions: basic concepts of microbial commensalism, colonization, infection, and disease." *Infection and Immunity* 68(12): 6511-6518.

Claverie J.M., C. N. (2006). *Bioinformatics for Dummies*, Wiley Publishing.

Dai X, Zhuang. Z., Zhao PX (2010). "Computational analysis of miRNA targets in plants: current status and challenges." *Briefings in Bioinformatics* 12(2): 115-121.

de-Jong H., Pietersma. H., Cordes M., Kuipers O.P., Kok J., (2012). "PePPER: a webserver for prediction of prokaryote promoter elements and regulons." *BMC Genomics* 13(1): 299.

Fones H., Davis. C. A., Rico A., Fang F., Smith J.A., Preston G.M., (2010). "Metal hyperaccumulation armors plants against disease." *PLoS Pathog* 6(9): 1.

Franza T., Mahe. B., Expert D., (2005). "Erwinia chrysanthemi requires a second iron transport route dependent of the siderophore achrophore achromobactin for extracellular growth and plant infection." *Mol Microbiol* 55: 261-275.

Griffiths-Jones S., Saini. H. K., Dongen S.V., Enright A.J., (2008). "miRBase: tools for microRNA genomics." *Nucleic Acids Research* 36: D154-D158.

Gustafson A.M., Allen. E., Givan S., Smith D., Carrington J.C., Kasschau K.D., (2005). "ASRP: the Arabidopsis Small RNA Project Database." *Nucleic Acids Research* 33: D637-D640.

He F., Zhang. Y., Chen H., Zhang Z., Peng Y.L., (2008). "The prediction of protein-protein interaction networks in rice blast fungus." *BMC Genomics* 9: 519.

Flor HH. (1971). "Current status of the gene-for-gene concept." *Annu Rev Phytopathol* 9: 275-296.

Hsiao Y.M., Liao. H. Y., Lee M.C., Yang T.C., Tseng Y.H., (2005). "Clp upregulates transcription of engA gene encoding a virulence factor in xanthomonas campestris by direct binding to the upstream tandem Clp sites." *Febs Lett* 579: 3525-3533.

Huang D.W., Sherman. B. T., Lempicki R.A., (2009). "Systematic and integrative analysis of large gene lists using DAVID bioinformatics resources." *Nat Protoc* 4: 44–57.

Ito T., Chiba. T., Ozawa R., Yoshida M., Hattori M., Sakaki Y., (2001). "A comprehensive two-hybrid analysis to explore the yeast protein interactome." *P Natl Acad Sci USA* 98(8): 4569-4574.

Ito T., T. K., Muta S., Ozawa R., Chiba T., Nishizawa M., Yamamoto K., Kuhara S., Sakaki Y., (2000). "Toward a protein-protein interaction map of the budding yeast: a comprehensive system to examine two-hybrid interactions in all possible combinations between the yeast proteins." *P Natl Acad Sci USA* 97(3): 1143-1147.

Jehl M.A., A. R., Rattei T., (2011). "Effective-a database of predicted secreted bacterial proteins." *Nucleic Acids Research* 39: D591-595.

Jiang B.L., L. J., Chen L.F., Ge Y.Y., Hang X.H., He Y.Q., Tang D.J., Lu G.T., Tang J.L., (2008). "DsbB is required for the pathogenesis process of xantomonas campestris pv. campestris." *Mol Plant Microbe In* 21(8): 1036-1045

Jonsson P., C. T., Zicha D., Bates P., (2006). "Cluster analysis of networks generated through homology: automatic identification of important protein communities involved in cancer metastasis." *BMC Bioinformatics* 7: 2.

Kim S.K., N. J. W., Rhree J.K., Lee W.J., Zhang B.T., (2006). "miTarget: microRNA target gene prediction using a support vector machine." *Bioinformatics* 7(1): 441.

Kurubanjerdjit N., Huang C.H., Lee Y.L., Tsai Jeffrey J.P, Ng K.L. (2013a). "Prediction of microRNA-regulated protein interaction pathway in Arabidopsis using machine learning algorithms," *Computers in Biology and Medicine*, 43(11), 1645-1652.

Kurubanjerdjit N., Tsai Jeffrey J.P, Sheu C.Y., Ng K.L. (2013b). "The prediction of protein-protein interaction of A. thaliana and X. campestris pv. campestris based on protein domain and interolog approaches", *Plant OMICS*, 6(6), 388-398.

Kurubanjerdjit N., Tsai Jeffrey J.P, Huang C.H., Ng K.L. (2014). Disturbance of A. thaliana microRNA-regulated pathways by Xcc bacterial effector proteins, *Amino Acids* 46(4), pp. 953-961.

Li Z.G., H. F., Zhang Z., Peng Y.L., (2011). "Prediction of protein-protein interactions between ralstonia solanacearum and arabidopsis thaliana." *Amino Acids* 42(6): 2363-2371.

Lin N., W. B., Jansen R., Gerstein M., Zhao H., (2004). "Information asssesment on predicting protein-protein interactions." *BMC Bioinformatics* 5: 154.

Lower M., S. G. (2009). "Prediction of type III secretion signals in genomes of gram-negative bacteria." *PLoS One* 4(6): 1.

Mandoli D.F., O. R. (2000). "The importance of emerging model systems in plant biology." *J Plant Growth Regul* 19(3): 249-252

Matys V, K.-M. O., Fricke E, Liebich I, Land S, Barre-Dirrie A, Reuter I, Chekmenev D, Krull M, Hornischer K, Voss N, Stegmaier P, Lewicki-Potapov B, Saxel H, Kel AE, Wingender E., (2006). "TRANSFAC and its module TRANSCompel: transcriptional gene regulation in eukaryotes." *Nucleic Acids Research* 34: D108-110.

Meyer D., L. E., Roby D., Arlat M., Kroj T., (2005). "Optimization of pathogenicity assays to study the arabidopsis thaliana–xanthomonas campestris pv. campestris pathosystem." *Mol Plant Pathol* 6(3): 327-333.

Morgan T.D., B. P., Kramer K.J., Basibuyuk H.H., Quicke D.L.J., (2002). "Metals in mandibles of stored product insects: do zinc and manganese enhance the ability of larvae to infest seeds?" *J Stored Prod Res* 39: 65-75.

Norambuena T., M. F. (2010). "The Protein-DNA Interface database." *BMC Bioinformatics* 11: 262.

Palla G., D. I., Farkas I., Vicsek T., (2005). "Uncovering the overlapping community structure of complex networks in nature and society." *Nature* 435: 814-818.

Pinzon A., R.-R. L. M., Gonzalez A., Bernal A., Restrepo S., (2010). "Targeted metabolic reconstruction: a novel approach for the characterization of plant-pathogen interactions." *Brief Bioinform* 12(2): 151-162.

Rhee SY, B. W., Berardini TZ, Chen G, Dixon D, Doyle A, Garcia-Hernandez M, Huala E, Lander G, Montoya M, Miller N, Mueller LA, Mundodi S, Reiser L, Tacklind J, Weems DC, Wu Y, Xu I, Yoo D, Yoon J, Zhang P., (2003). "The arabidopsis information resource (TAIR): a model organism database providing a centralized, curated gateway to arabidopsis biology, research materials and community." *Nucleic Acids Research* 31(1): 224-228.

Rolke Y., L. S., Quidde T., Williamson B., Schouten A., Weltring K.M., Siewers V., Tenberge K.B., Tudzynski B., Tudzynski P., (2004). "Functional analysis of H2O2-generating systems in Botrytis cinerea: the major Cu-Zn-superoxide dismutase (BCSOD 1) contributes to virulence on French bean, whereas a glucose oxidase (BCGOD1) is dispensable." *Mol Plant Pathol* 5: 17-27.

Sanseverino W., R. G., De-Simone M., Faino L., Melito M., Stupka E., Frusciante L., Ercolano M.R., (2010). "PRGdb: a bioinformatics platform for plant resistance gene analysis." *Nucleic Acids Research* 38: D814–D821.

Stahl E.A., B. J. G. (2000). "Plant-pathogen arms races at the molecular level." *Curr Opin Plant Biol* 3(4): 299-304.

Tang D.J., L. X. J., He Y.Q., Feng J.X., Chen B., Tang J.L., "The zinc uptake regulator Zur is essential for the full virulence of Xanthomonas campestris pv campestris." *Mol Plant Microbe Interact* 18: 652-658.

Tsao T.H., C. C. H., Huang Chi-Yang F., Lee S.A., (2011). Systems and computational biology-molecular and cellular experimental systems. In: Prof.Ning-Sun Yang (ed) *The prediction and Analysis of Inter- and Intra-Species Protein-Protein Interaction.* China, InTech.

Tsuji J., S. S. C. (1988). "Xanthomonas campestris pv. campestris induced chlorosis in Arabidopsis thaliana." *Arabidopsis Information Service* 26: 1-8.

Tsuji J., S. S. C. (1992). "First report of the natural infection of arabidopsis thaliana by xanthomonas campestris pv. campestris." *Plant Dis* 76: 539.

Tsuji J., S. S. C., Hammerschmidt R., (1991). "Identification of a gene in arabidopsis thaliana that controls resistance to xanthomonas campestris pv. campestris." *Physiol Mol Plant P* 38: 57-65.

Tucker S.L., T. T. R., Tasker K., Jacob C., Giles G.,Egan M.,Talbota N.J., (2004). "A fungal metallothionein is required for pathogenicity of Mangaporthe grisea." *Plant Cell* 16: 1575-1588.

Uetz P., G. L., Cagney G., Mansfield T.A., Judson R.S., Knight J.R., Lockshon D., Narayan V., Srinivasan M., Pochart P., Qureshi-Emili A., Li Y., Godwin B., Conover D., Kalbfleisch T., Vijayadamodar G., Yang M., Johnston M., Fields S., Rothberg J.M., (2000). "A comprehensive analysis of protein-protein interactions in Saccharomyces cerevisiae." *Nature* 403(6770): 623-627

Yu H., L. N. M., Lu H.X., Zhu X., Xia Y., Han J.D., Bertin N., Chung S., Vidal M., Gerstein M., (2004). "Annotation transfer between genomes: protein-protein interologs and protein-dna regulogs." *Genome Research* 14(6): 1107-1118.

Zhang Z., Y. J., Li D., Zhang Z., Liu F., Zhou X., Wang T., Ling Y., Su Z., (2010). "PMRD: Plant microRNA database. Nucleic Acids Research." *Nucleic Acids Research* 38: D806-D813.

Chapter 7

Computational modelling of the Alu-carrying RNA network in Th17-mediated autoimmune diseases

Kung-Hao Liang
Medical Research Department
Taipei Verterans General Hospital

Abstract

In the human genome, 10% of the nucleotide sequences were Alus which were retrotransposon-produced genomic repeats. Despite occasional evidence of Alu-induced genetic diseases, it remained a mystery whether these Alu elements play substantial physiological roles comparable with its proportion in the human genome. Recently, cytosolic sense- and antisense-Alu carrying mature RNAs, corresponding to more than 1300 protein coding and various non-coding genes, were shown to form a network of mutual regulations. Messenger RNA transcripts of genes pertinent to the immunological Th17 maturation, including CCL5, CCR6, IL23R, IL2RA, IL1R1, CD28 and REL, consistently carry Alu in the sense direction. On the other hand, other immunological genes such as CXCL16, IFNAR2, CD302, CDH1, IL28RA (a.k.a. IFNLR1) and JAK3, all carry Alus only in the antisense direction. The Alu sequences facilitated RNA-RNA interactions, resulting in a RNA regulatory network which enabled a computational modelling of cellular state transitions. The transition from naïve T cells to mature Th17 cells was largely controlled by the relative transcriptional rates of genes carrying sense and antisense Alus, which showed an inherent inverse relationship of levels at equilibrium.

7.1 Background

The human immunological system coordinates reactions to environmental and residential micro-organisms in the human body. These reactions are broadly classified as Th1, Th2, Treg and Th17 reactions. Th17 is an adaptive, cellular immune response facilitated by the maturation of T helper(h)-17 cells [1]. In normal conditions, Th17 is activated upon pathogen invasions, particularly the extracellular bacteria and fungi [1]. Autoimmune diseases are chronic, debilitating diseases. Until now, pathogenic mechanisms of autoimmune diseases remain largely elusive, and the treatments have been suboptimal. Diseases such as systemic lupus erythematosus, multiple sclerosis, rheumatoid arthritis and psoriasis manifest as pathogenic inflammation in different body parts. In the meantime, all of them are ascribed to a common pathogenic mechanism, i.e. the attack of immune system to the body's own tissues. Recently, Th17 overreactions have been ascribed to various autoimmune diseases [2]. It is thus imperative to have a better understanding on the Th17 activation so as to formulate counteracting strategies for Th17-related autoimmune diseases.

Until now, the known molecular pathway of Th17 activation was rather complex. It comprised gene expression, cytokine stimulation, signal transduction and protein complex forming, which altogether involved a large number of genes interconnected in a way of intertwined positive and negative feedback loops. In such conditions, the dynamic behaviour become less intuitive. Regrettably, the molecular pathways still seemed incomplete, prohibiting a computational modelling of the dynamics from the cellular state A to state B.

The human genome encodes the blueprint of immune system. One tenth of the human genome is composed of a single class of non-coding elements, the short interspersed Alu repeats, which are only found in primate genomes [3, 4]. Recently, an unconventional thinking on Alu's regulatory role was proposed. Human messenger RNA transcripts carrying Alu elements in two opposite directions were demonstrated to form strong sense-antisense bindings dictated by the thermodynamic laws [5]. More than 1300 protein-coding genes carry sense or antisense Alu elements, mostly in the 3' untranslated region [5]. Such type of RNA duplex may trigger mutual regulation post-transcriptionally, offering a new angle of computational modelling. Additional molecular evidence of Alu mediated mRNA-mRNA binding was subsequently offered by an independent group [6]. This chapter will show how that the Alu mediated network can enable the modelling of state transition, for elucidating the dynamics behind state transitions.

7.2 Methods

7.2.1 Analysis of gene co-expressions upon Th17 activations

The microarray gene expression data were extracted from Gene Expression Omnibus (GSE42569) [2]. This dataset comprised four samples from two independent healthy human donors. Samples were either naïve T cells or Th17 cells activated by NaCl. Data have been normalised by submitters using the Mas 5.0 algorithm. We aimed to quantify the differences of mean RNA levels between the two states. As there are multiple probes designed to target one gene in this microarray platform (Affymetrix Human U133 Plus 2.0), the median value of signals from all probes ascribed to a gene were used to represent the gene level. This was done by the GSEA software (v2.0.10) [7].

7.2.2 Computational Modelling of the Alu-mediated network

A schematic diagram in Fig. 1 illustrates the major biological and biochemical reactions in the Alu-mediated network. Antisense- and sense-Alu carrying RNAs form double-stranded RNAs, which are then subject to the cutting of endonucleases. Shattered sense and antisense Alu transcripts then serve as guide strands in the RNA-induced silencing complex (RISC), which in turn suppress antisense- and sense-Alu carrying RNAs respectively.

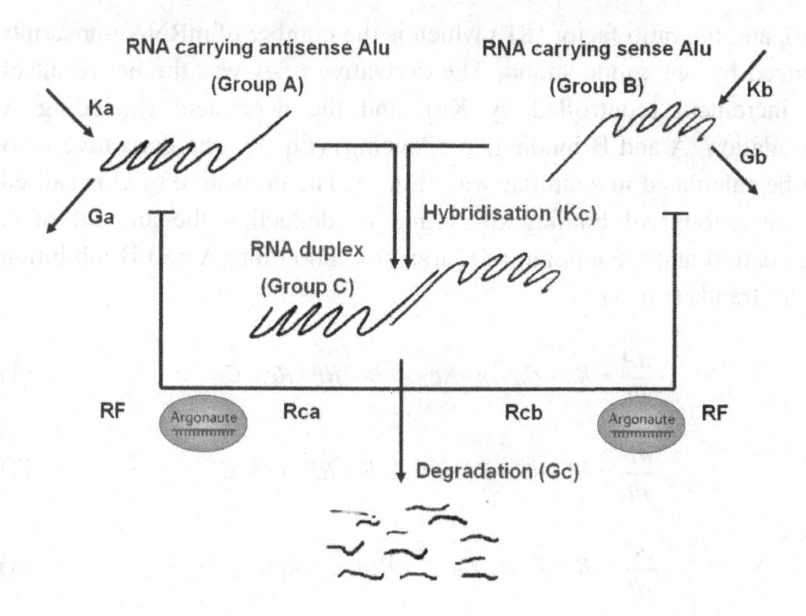

Fig. 1. A schematic diagram of the Alu mediated regulation network. RNA transcripts carrying antisense Alus (species A) and sense Alus (species B) hybridise and form a duplex structure (species C). Two biological properties were modelled, including the expression rates of the two species (Ka and Kb). Biochemical properties in this model include Kc (the hybridisation rate of A and B), Ga, Gb, Gc (degradation rates), Rca, Rcb (proportion of C which become guide strands for forming the RISC complex) and RF (number of silenced transcripts per a guide strand).

A set of mathematical equations were used to model the Alu-mediated RNA network. A and B indicate the time-dependent RNA levels of antisense- and sense-Alu carrying mRNAs respectively. The biological parameters of transcription rates of A and B species are shown by Ka and Kb respectively. The biochemical parameters include the degradation rates of A and B (shown by Ga and Gb respectively), the hybridisation rate of A and B (Kc), the degradation rate of C (Gc), the proportions of C to become A and B inhibition guide strands (Rca and

Rcb), and the ratio factor (RF) which is the number of mRNA transcripts silenced by one guide strand. The derivative of A was the net result of the increment (controlled by Ka), and the decrement (including A degradation, A and B binding, A silencing) (Eq. 1). The derivative of B can be calculated in a similar way (Eq. 2). The derivative of C is caused by the amount of binding of A and B, deducting the amount of C degradation and the amount of C transformation into A and B inhibition guide strands (Eq. 3).

$$\frac{dA}{dt} = Ka - Ga \cdot A - Kc \cdot A \cdot B - RF \cdot Rca \cdot C \tag{1}$$

$$\frac{dB}{dt} = Kb - Gb \cdot B - Kc \cdot A \cdot B - RF \cdot Rcb \cdot C \tag{2}$$

$$\frac{dC}{dt} = Kc \cdot A \cdot B - Gc \cdot C - Rca \cdot C - Rcb \cdot C \tag{3}$$

State transitions of Alu-carrying RNA were simulated using programming scripts running on the GNU Octave freeware (v.3.2.4).

7.3 Results

7.3.1 Sense-Alu carrying RNAs were co-activated during Th17 maturation

A high-quality gene expression profile of human Th17 maturation was recently published [2]. In this study, naïve CD4+ cells were harvested from healthy human subjects and then stimulated by a high concentration of NaCl to trigger an ex-vivo state transition into Th17 cells. Comparing the RNA levels of naïve cells and mature Th17 cells, CCL5, CCR6, IL23R, IL2RA, IL1R1, CD28 and REL were up-regulated (Table 1, Fig. 2). All of them carry Alus only in the sense direction. Particularly, IL23R and CCR6 have been shown to be the signature receptors of mature Th17 cells (Fischer, 2008). On the other hand, CXCL16, IFNAR2, CD302, CDH1, IL28RA (a.k.a. IFNLR1) and JAK3, which carry Alus only in the antisense direction, were all down-regulated (Table 1, Fig. 2). The data suggested that in parallel to the process of Th17 maturation, the balance of gene levels tilted, and the direction of gene expression changes correlates with the direction of the Alus in the messenger RNAs.

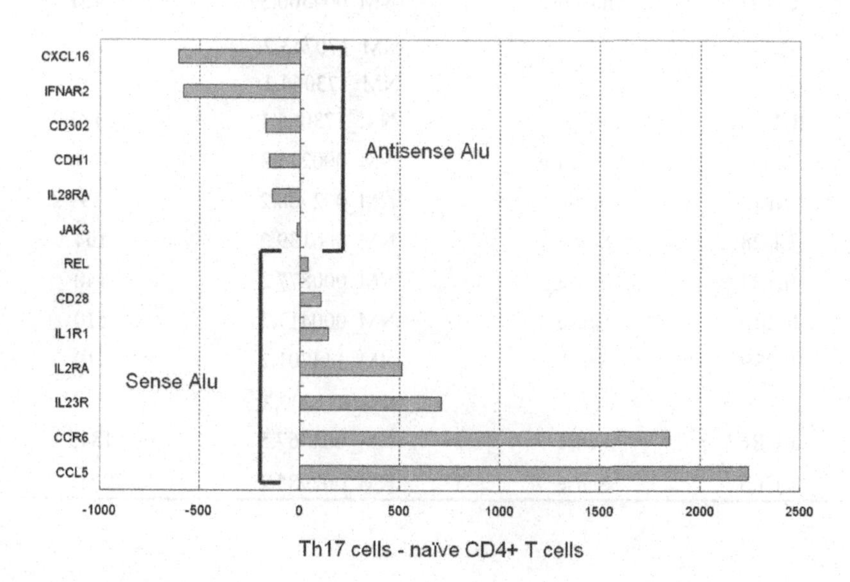

Fig. 2. Differences of RNA levels between Th17 cells and naïve CD4+ T cells. Immunological genes tagged by sense-Alu were up-regulated while those tagged by antisense Alu were down-regulated during the Th17 state transition.

Table 1. Levels of sense and antisense Alu carrying mRNAs in response to Th17 maturation.

Gene Symbol	Alu directions	Refseq accession number	RNA level difference
CXCL16	Antisense	NM_022059.2	−607
IFNAR2	Antisense	NM_207585.1	−581
CD302	Antisense	NM_014880.4	−170

(*Continued*)

Table 1. (*Continued*)

Gene Symbol	Alu directions	Refseq accession number	RNA level difference
CDH1	Antisense	NM_004360.3	−151
IL28RA	Antisense	NM_170743.2; NM_173064.1; NM_173065.1	−137
JAK3	Antisense	NM_000215.3	−13
REL	Sense	NM_002908.2	39
CD28	Sense	NM_006139.2	104
IL1R1	Sense	NM_000877.2	140
IL2RA	Sense	NM_000417.2	510
IL23R	Sense	NM_144701.2	710
CCR6	Sense	NM_031409.3; NM_004367.5	1845
CCL5	Sense	NM_002985.2	2241

7.3.2 Regulation of sense- and antisense-Alu carrying genes at equilibrium

Th17 maturation can be seen as a state transition process where naïve T cells and mature Th17 cells are the two cellular steady states. Since the major alteration of gene levels underlying the state transition was consistent with the Alu directions, the state transition can be modelled by the Alu-mediated network (see Methods). The two steady states occur when levels of A, B and C are at equilibrium and their derivatives are zero (i.e. no change of RNA levels). These steady states are controlled by transcription rates Ka and Kb. Holding the biochemical parameters as constants and adjusting the cellular transcription rate Kb/Ka in various values between [1/17, 17] a parabolic curve of steady states emerges (Fig. 3). A typical inverse relationship was found between A and B. An up-regulation of B during Th17 maturation coincides with a down-regulation of A, which was largely controlled by the ratio of Kb/Ka.

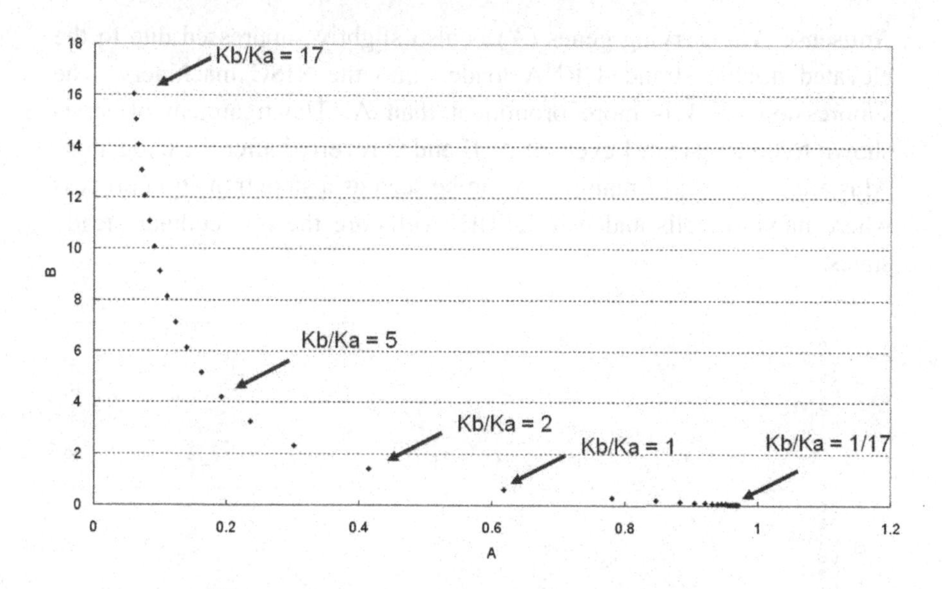

Fig. 3. An inherent inverse relationship of A and B levels at equilibrium. The levels are determined when the biochemical parameters are constant (Ga=Gb=3.5; Kc=1; Gc=0.5; Rca=0.25; Rcb=0.25). The transcription rate of A is also kept constant (Ka=3.5). Different Kb are given. The activation of species B (controlled by the parameter of Kb) is accompanied by a decrease of species A.

7.3.3 Antisense RNA interference offers a transient therapeutic effect

We then moved on to see if Kb/Ka was not changed, what will happen if exogenous RNA were given. As the Th17 related genes CCL5, CCR6, IL23R, IL2RA, IL1R1, CD28 and REL all carry sense Alus, exogenous antisense Alu RNA may concurrently suppress these genes. The time-course levels of A, B and C upon one treatment of exogenous antisense Alu (at the time point #5) is shown in Fig. 4. Levels of A, B and C are stable before the treatment due to a fixed Ka/Kb ratio of 1. A maximum peak of exogenous antisense Alu occurs at the time point #6. An increase of the double stranded RNA (C) and a decrease of the level of B is due to the binding of exogenous nucleotide with sense Alu carrying genes (B).

Antisense Alu carrying genes (A) is also slightly suppressed due to the elevated double stranded RNA loaded into the RISC machinery. The suppression of B is more prominent than A. The treatment effect is shown to be transient. Levels of A, B and C reverted after the exogenous Alus are exhausted17 maturation can be seen as a state transition process where naïve T cells and mature Th17 cells are the two cellular steady states.

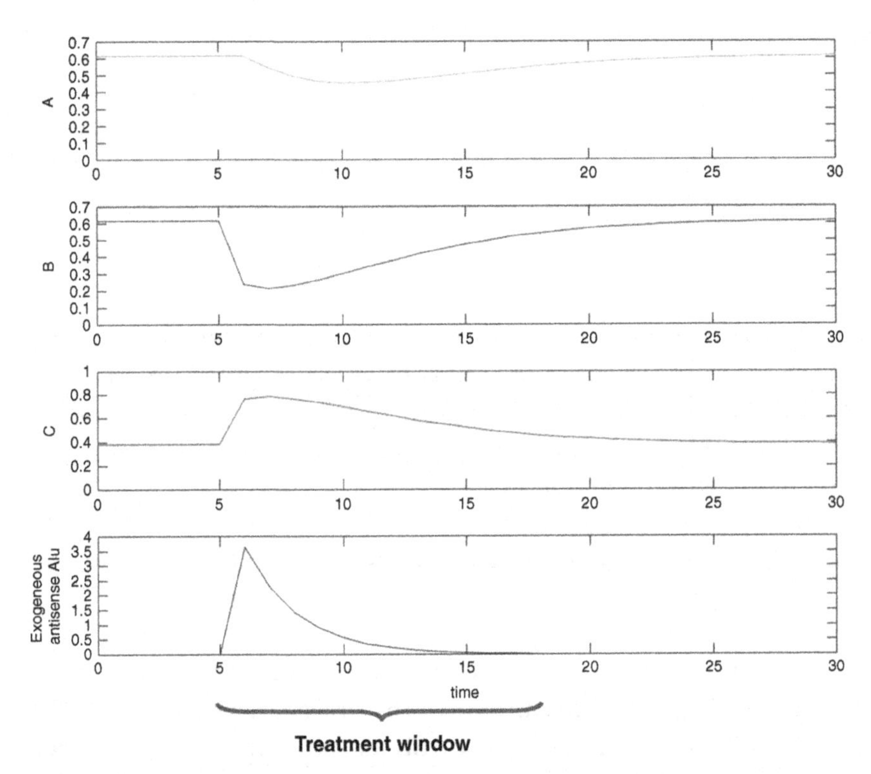

Fig. 4. Dynamic responses of species A, B and C when a single treatment of exogenous antisense Alu was given at time point #5. Level of species A dropped slightly and then returned back to the original level. Level of species B dropped more prominently and then returned back to the original level. Level of C increased prominently due to the binding of species B and the exogenous antisense Alu, then returned to the original level. Exogeneous antisense Alu level reached the peak at the time point #6, then were gradually exhausted. The parameters were identical to those in Fig. 3, except Kb was set constantly at 3.5.

To sustain the treatment effect, repeated treatment may be required. Figure 5 shows the time-course levels of A, B and C when the treatment is repetitively given every 5 time unit. This way, the level of group B is kept low, while group A is also slightly reduced.

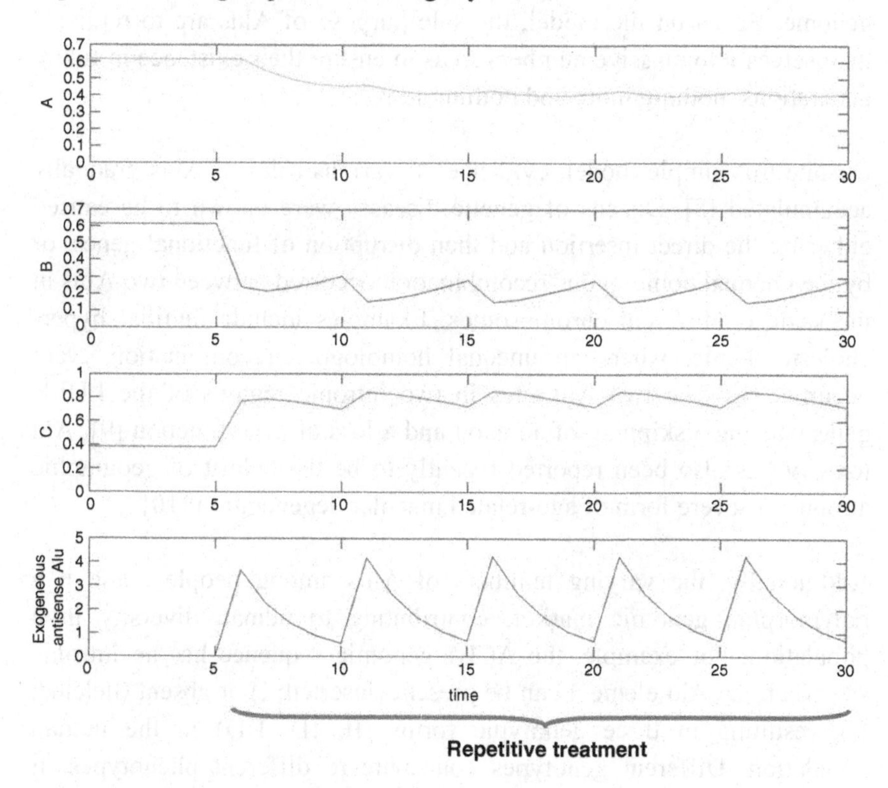

Fig. 5. Transient responses of species A, B and C when repetitively treatment of exogenous antisense Alu was given every 5 hours. Parameters were the same as in Fig. 4.

7.4 Discussion

During the course of evolution, human genomic Alu repeats were replicated by retrotransposition to more than one million copies [3]. Each Alu element is ~300 nucleotide bases. It is generally believed that the retrotransposition machinery is partly borrowed from L1, a ~6000 base protein-coding retrotransposon [8]. The vast number yet non-coding nature of Alu makes us wonder whether the non-coding elements play constructive or destructive role in human physiology? The selfish gene model is commonly cited to explain the current role of Alus in the human genome. Based on the model, the sole purpose of Alus are to replicate themselves into massive numbers so as to ensure their existence in future generations, nothing more and nothing less.

Despite this simple model, evidence on various roles of Alus gradually accumulated [8]. Dozens of genetic diseases were known to be caused either by the direct insertion and then disruption of functional genes, or by the unequal homologous recombination occurred between two Alus in the same or different chromosomes. Examples include familial hyper-cholesterolemia, where an unequal homologous recombination event occurred between two Alu sites in two intronic regions of the LDLR gene, causing a skipping of an exon and a loss of gene function [9]. Alu toxicity has also been reported recently to be the culprit of geographic atrophy, a severe form of age-related macular degeneration [10].

Additionally, the varying numbers of Alus among people made it a polymorphic genomic marker, contributing to human diversity in a population. For example, the ACE1 genomic sequence has an intronic site where an Alu element can be present (inserted; I) or absent (deleted, D), resulting in three genotypic forms (II, ID, DD) in the human population. Different genotypes contribute to different phenotypes. It was reported that the DD type correlated with higher serum activity of ACE1, followed by ID and II genotypes.

It was also postulated that Alus have offered a mutational mechanism during the past history of human evolution, mainly through the homologous recombination of Alu elements in different chromosomal locations [8].

In this chapter, a new strategy of computational modelling was explored, based on the shared sense Alu elements in major genes involved in the

Th17 activation. Transcriptomic microarray data showed that during Th17 maturation, sense Alu-carrying genes were up-regulated, while antisense Alu-carrying RNAs were down-regulated. A transient RNA interference approach can produce the tilt of balance of Th17 genes. Since this method does not change the biological transcription rates, the original state can be reverted after the exogenous RNA was stopped.

In conclusion, we have shown that the up and down regulation of immunological genes correlated with the directions of Alus in their mRNA. This knowledge can be used for the computational modelling of state transitions, which will be otherwise very difficult to model quantitatively.

References

[1] A. Fischer, "Human immunodeficiency: Connecting STAT3, Th17 and human mucosal immunity," Immunology and Cell Biology, vol. 86, pp. 549-551, 2008/07/22 2008.

[2] M. Kleinewietfeld, A. Manzel, J. Titze, H. Kvakan, N. Yosef, R. A. Linker, *et al.*, "Sodium chloride drives autoimmune disease by the induction of pathogenic TH17 cells," Nature, vol. 496, pp. 518-522, 2013/03/06 2013.

[3] M. A. Batzer and P. L. Deininger, "ALU REPEATS AND HUMAN GENOMIC DIVERSITY," Nature Reviews Genetics, vol. 3, pp. 370-379, 2002/05/01 2002.

[4] E. S. Lander, L. M. Linton, B. Birren, C. Nusbaum, M. C. Zody, J. Baldwin, *et al.*, "Initial sequencing and analysis of the human genome," Nature, vol. 409, pp. 860-921, 2001/02/15 2001.

[5] K.-H. Liang and C.-T. Yeh, "A gene expression restriction network mediated by sense and antisense Alu sequences located on protein-coding messenger RNAs," BMC Genomics, vol. 14, p. 325, 2013.

[6] C. Gong and L. E. Maquat, "lncRNAs transactivate STAU1-mediated mRNA decay by duplexing with 3 UTRs via Alu elements," Nature, vol. 470, pp. 284-288, 2011/02/10 2011.

[7] A. Subramanian, P. Tamayo, V. K. Mootha, S. Mukherjee, B. L. Ebert, M. A. Gillette, *et al.*, "Gene set enrichment analysis: A knowledge-based approach for interpreting genome-wide expression profiles," Proceedings of the National Academy of Sciences, vol. 102, pp. 15545-15550, 2005/09/30 2005.

[8] R. Cordaux and M. A. Batzer, "The impact of retrotransposons on human genome evolution," Nature Reviews Genetics, vol. 10, pp. 691-703, 2009/10 2009.

[9] H. H. Hobbs, M. A. Lehrman, T. Yamamoto, and D. W. Russell, "Polymorphism and evolution of Alu sequences in the human low density lipoprotein receptor gene," Proceedings of the National Academy of Sciences, vol. 82, pp. 7651-7655, 1985/11/01 1985.

[10] H. Kaneko, S. Dridi, V. Tarallo, B. D. Gelfand, B. J. Fowler, W. G. Cho, *et al.*, "DICER1 deficit induces Alu RNA toxicity in age-related macular degeneration," Nature, vol. 471, pp. 325-330, 2011/02/06 2011.

Chapter 8

Principal component analysis based unsupervised feature extraction applied to bioinformatics analysis

Y-h. Taguchi

Department of Physics, Chuo University, Tokyo 112-8551, Japan

Mitsuo Iwadate

Departmet of Biological Sciences, Chuo University, Tokyo 112-8551, Japan

Hideaki Umeyama

The School of Pharmacy, Kitasato University, Tokyo 108-8641, Japan

Yoshiki Murakami

Department of Hepatology, Osaka City University Graduate School of Medicine, Osaka, Japan

8.1 Introduction

In bioinformatics analysis, it is very usual that there are more features than samples. You are supposed to analyze gene expression profile composed of tens of thousands of genes and less than a hundred samples. In this case, it is unrealistic to assume that all genes contribute to something that you would like to investigate. "something" can be either disease, reaction toward some drugs, or anything else. Then, you are usually willing to identify limited number of genes that truly contribute to something of your interest. But, how? It is very natural to do this by selecting genes that can successfully discriminate samples of interest from those supposed to be control. Or you can simply select genes highly expressive or suppressive in samples of interest than in control ones.

However, this strategy might not give you an appropriate set of genes because class labels may not always be true. For example, target samples include more females than control ones, or more aged patients are in disease samples than in healthy controls. In that case, any successful discrimination or over/under expression may not be because of something of your interest, but not intended unbalance of something without your interest between sampled interest and control ones. One may think that it is not a problem at all, since you can select equal number of males/females between two classes, or you can match mean age between two classes. However, this does not also solve the problem completely. Ratio of smokers or frequent drinkers may not be same between two classes. Practically, it is impossible to prepare samples where all features but those you are interested in obey same distribution between two classes.

Unsupervised methodologies may solve this difficulty, since, in contrast to the supervised methodologies, it can classify (or to find clusters of) samples without accessing class labels. Although it is reasonable to be afraid of less ability of unsupervised classifications because of their possible vulnerability to noise, instead of that, unsupervised methods are less likely affected by mislabeling. As for the above examples associated with unintentionally unbalanced distribution of not focused features, primary cluster (classification) will be not due to the labeling of interest but coincident with the unfocused classification, e.g., aging or sex. Then, we can have the opportunity to notice that samples are highly unbalanced with some features not focused. Or if you find no clustering (classification) is coincident with the labeling of interest, you can have opportunity to terminate study and go back to the beginning of project to prepare updated samples with balanced features.

However, this strategy is highly opposed to the recent trends that encourage to find some criteria to classify samples associated with given labeling. These criteria was usually implemented in some model. These so-called model-based approach is usually powerful to exclude some features not related to the discrimination between targets and control samples and to identify critical features necessary to classify target samples from control samples. However, as mentioned in the beginning of this chapter, in bioinformatics analysis, it is very usual that we cannot have enough samples to train models so as to classify two classes well. In that case, we are not supposed to get limited number of features that classify two classes using model based approaches. Another difficulty of model based approaches raises when there are more classes than two without the preknowledge about the relationship among multiple classes. For example, suppose that we ought to treat time sequence data. In this case, each time point corresponds to each class, among which no information about in which time point genes are expressive/depressive. Of course, although it is always possible to identify features that are not constant over all time points, e.g., using ANOVA, it sometimes cannot provide us enough information, since time dependence, e.g., in which time point which genes are expressive/repressive, which cannot be obtained by ANNOVA is important. Although another strategy is to use clustering, successful usages of clustering analysis often requires to identify in advance limited number of features which are coincident with multiple classes clustering. Since identification of features coincident with multiple categorical classes is nothing but the purpose of study, the usage of clustering in order to identify features is not a realistic strategy.

In this chapter, we will demonstrate that principal component analysis is useful to identify limited number of critical genes when samples are much less than the total number of genes. We will review various applications of this methodology achieved since the publication of the previous review [Taguchi *et al.* (2015c)].

8.2 Previous researches using PCA for gene selection

Before discussing our methodology, we briefly review studies that make use of PCA for gene selection. The most popular strategies is to identify limited number of features to preserve primary structure of PCA [Wang and Gehan (2005); Krzanowski (1987)]. In this regard, PCA is not a tool but rather a purpose. This is opposed to our strategy that often identifies features based upon miner PCs that contribute less (see the following application examples). Thus, it principally differs from our methodology in spite of apparent similarities.

Alternatively, some studies make use of PCA to identify genes. Jonnalagadda and Srinivasan [Jonnalagadda and Srinivasan (2008)] tried to identify differentially expressed genes based upon contribution to PCs, which is very similar to our methodology. The potential difference between ours and theirs is that they have evaluated difference of contributions between two classes. Our methodology, as can be seen below, we never compute difference between two classes directly, but identify gene as outliers along specified PCs. We also used difference between two classes in order to identify PCs used for outlier identification and P-values are evaluated assuming χ^2 distributions, in spite of apparent similarity between ours and theirs, there are principal differences. First of all, since they performed pairwise comparisons, extension towards categorical multiclass is unclear, although it is straightforward in our methodology. Second, after identifying PCs used for outlier identification, we do not directly require individual genes to be distinct between two classes. Thus, our methodology is more data driven and has more opportunity to identify more feasible genes.

To our knowledge, there are many apparently similar methodologies, ours are outstanding from other similar methodology. At least, no other previous methodologies could not be applied to wide range of problems to which PCA based unsupervised FE can be successfully applied.

8.3 Principal component analysis based unsupervised feature extraction

In this section, we explain the basic procedure to apply the principal component analysis (PCA) based unsupervised feature extraction (FE) to gene expression/promoter methylation profiles.

8.3.1 *Samples embedding versus genes embedding*

Suppose that x_{ij} is gene expression/promoter methylation of ith gene and jth sample. The matrix X is suppose to have x_{ij} as its element. In the usual usage of PCA, samples are embedded into low dimensional space. Coordinates in the embedded space was called as principal component (PC) scores and obtained as eigen vectors, $\boldsymbol{u}_k$, of the covariance matrix $X^T X$

$$X^T X \boldsymbol{u}_k = \lambda_k \boldsymbol{u}_k$$

where $\boldsymbol{u}_k$ and λ_k are kth eigen vector and eigen value. u_{kj} corresponds to PC scores of jth sample. Then, PC loadings attributed to each gene can be obtained as ith component of the vector $\boldsymbol{v}_k = X\boldsymbol{u}_k$. In the above, X is assumed to be normalized as $\sum_j x_{ij} = 0$ and $\sum_j x_{ij}^2 = M$ where M is the total number of samples.

In contrast to the above standard usage of PCA, in PCA based unsupervised FE, not samples but genes are embedded into low dimensional space. Then, PC scores attributed to each genes are computed as eigen vectors of the Gram matrix XX^T,

$$XX^T \boldsymbol{u}_k = \lambda_k \boldsymbol{u}_k$$

and PC loadings attributed to each sample is given as vector $\boldsymbol{v}_k = X^T \boldsymbol{u}_k$. Here the normalization differs from the above and $\sum_i x_{ij} = 0$ and $\sum_i x_{ij}^2 = N$ where N is the total number of genes. Thus, hereafter, PC scores are attributed to genes and PC loadings attributed to samples. The number of eigen vector and eigen values are smaller number among N and M, although in the followings $M < N$ thus the number of eigen values/vectors are M.

Although one may wonder if these two PCA, sample embedding and gene embedding, are equivalent, these two differ from each other because of the distinct mean extraction $\sum_i x_{ij} = 0$ or $\sum_j x_{ij} = 0$. Since PCA is the diagonalization of the product of X, the effect of distinct mean extraction is non-liner. Therefore generally there are noways to infer the results of gene embedding from those of sample embedding and vise versa.

8.3.2 *Identification of critical genes as outliers*

In order to identify genes critical to something of interest, we need to specify PC loadings used for outlier identification (see below). There are multiple strategies that identify PC loadings.

- **Distinct expression between target samples and control samples**: The most simplest an simple way is to identify PC loadings, $\boldsymbol{v}_k$s, distinct between target samples and control samples.
- **Coincidence**: When PCA based unsupervised FE was used for integrating matched data sets, coincidence between PC loadings, $\boldsymbol{v}_k$s, between matched data set. For example, when microRNA (miRNA) and mRNA expression are integrated, pairs of PC loadings of miRNA and mRNA associated with negative correlations are employed.
- **Others** Dependent upon the purpose of study, some other criteria to identify PC loadings, $\boldsymbol{v}_k$s, used for outliers identification is possible. For example, for identifying genes associated with cell division cycles, pairs of PC loadings that form limited cycles are employed (see below).

After PC loadings are identified, outlier genes are selected assuming multiple normal distribution for PC scores, $\boldsymbol{u}_k$s, attributed to each genes. P-values are computed using χ^2 distribution,

$$P\left[> \sum_k \left(\frac{u_{ki}}{\sigma_k}\right)^2\right]$$

where $P[> x]$ is the cumulative probability that argument is larger than x under χ^2 distribution and σ_k is the standard deviation of u_{ki}. The summation is taken over PCs selected for the selecting outliers.

Obtained P-values are further adjusted using BH criterion [Benjamini and Hochberg (1995)] and genes associated with small enough adjusted P-values are selected (Typically, adjusted P-values less than 0.01).

8.4 Identification of miRNA-mRNA interaction

The first application of PCA based unsupervised FE discussed in this chapter is identification of miRNA-mRNA interactions. microRNAs are small non-coding RNAs whose canonical function is known to be translation inhibition/mRNA degradation. Potential difficulty of miRNA-mRNA interaction identification is too many possibilities. The number of known miRNAs is more than one thousand while mRNAs are as many as tens of thousand. Since mRNAs are targets of miRNAs if their 3' UTRs match with miRNA seed sequence of eight bases, millions of pairs are possible. Experimentally, it is unrealistic to check all of possible pairs in each experiment. Although various bioinformatics inference was proposed, since they are primarily sequence based, it is basically impossible to consider context dependence of miRNA-mRNA interaction.

Then, the present state of art methodology is to pre-screen mRNAs/miRNAs associated with significant differential expression between target samples and control samples. This usually effectively reduces the number of possible interaction by reducing number of miRNAs/mRNAs considered.

In this subsection, we demonstrated that PCA based unsupervised FE can successfully identify candidate mRNAs/miRNAs for miRNA-miRNA interaction identification.

8.4.1 *miRNA-mRNA interaction in various cancers*

Among various known functions of miRNAs, its contribution to tumor genesis attracts broad interest [Jansson and Lund (2012)]. This is because miRNAs can be both therapy and diagnosis targets. Then, many tried to identify miRNA-mRNA interaction in cancers, too. In spite of the researchers' general interest in cancer genesis, miRNA-mRNA interaction was usually identified in individual cancer. To our knowledge, there are no single studies that report miRNA-mRNA interaction for multiple cancers. The reason of this missing reports is possibly because of the lack of common criterion to pre-screen miRNAs/mRNAs associated with significant differential expression between patients and healthy controls. Table 8.1 shows the variety of

Table 8.1 Criteria to identify miRNAs/mRNAs associated with significant differential expression between patients and healthy controls.

Cancers	Significance criteria: miRNA	Significance criteria: mRNA	References
HCC	FDR ≤ 0.01; $\log_2$ FC ≥ 1		[Ding *et al.* (2015)]
NSCLC	FDR < 0.1 by SAM		[Ma *et al.* (2011)]
ESCC	literature searches	FC > 1.5	[Wu *et al.* (2013)]
	FDR < 0.05	FC > 3; FDR < 0.001	[Yang *et al.* (2015)]
	FDR < 0.05		[Meng *et al.* (2014)]
PC	no description		[Zhang *et al.* (2012)]
CRC	FDR < 0.05		[Fu *et al.* (2012)]
CC	FC > 1.2; FDR< 0.1		[Li *et al.* (2011)]
BC	miRtest [Artmann *et al.* (2012)]	no description	[Bleckmann *et al.* (2015)]
PDA	FDR* < 0.05; $\mid \log \text{FC} \mid > 1$		[Liu *et al.* (2015)]

HCC: Hepatocellular carcinoma, NSCLC: non-small cell lung cancer, ESCC: esophageal squamous cell carcinoma, PC: prostate cancer, CRC: colorectal cancer, CC: colon cancer, BC: Breast cancer, PDA: Pancreatic cancer, FC: fold change.
*Bonferroni's correction-adjusted P-value.

criteria used for pre-screening mRNAs/miRNAs in various cancers within the recent five years. It is obvious that none paid attention on the diversity of criteria. In some sense, it is natural since the researchers in each study are interested in a specific cancer studied in each study.

One may wonder why it is problematic if obtained outcomes are biologically feasible (and of course, it was so, since otherwise the papers could not be published). The problem is, if there are no consensus on how we pre-screen miRNAs/mRNAs prior to the identification of miRNA-mRNA interactions, that people may try to seek successful criteria until they succeed. Then, seeking successful pre-screening criteria may become like optimization. This may bias the outcomes (Or in other words, many false positives are reported). Thus, if possible, a single criterion applicable to multiple cancers is desirable.

One of difficulties of single criterion applicable to multiple studies is the dependence of P-values upon the number of samples. As samples increase, less differences becomes feasible. In this case, we have to employ more strict criteria in order to reduce the number of mRNAs/miRNAs that pass the pre-screening. In this sense, PCA based unsupervised FE is promising, since it employs P-values attributed to genes whose numbers do not vary. Thus, we do not have to tune the pre-screening criteria dependent upon sample numbers. Recently, I applied PCA based unsupervised FE in order to pre-screen miRNAs/mRNAs in various cancers [Taguchi (2016a)] and successfully identified reliable sets of miRNA-mRNA pairs for multiple cancers using a single criterion. Figure 8.1 shows how to identify miRNA-mRNA interactions using PCA based unsupervised FE. First, mRNA/miRNA expression profiles was pre-screened separately using PCA based unsupervised FE. Then, mRNAs/miRNAs expressed significantly and distinctly between tumors and normal tissues were further screened. Finally, among those pre-screened, miRNA-mRNA interactions were identified using TargetScan [Agarwal *et al.* (2015)], which is one of the most trustable sequence based miRNA-mRNA interaction inference databases.

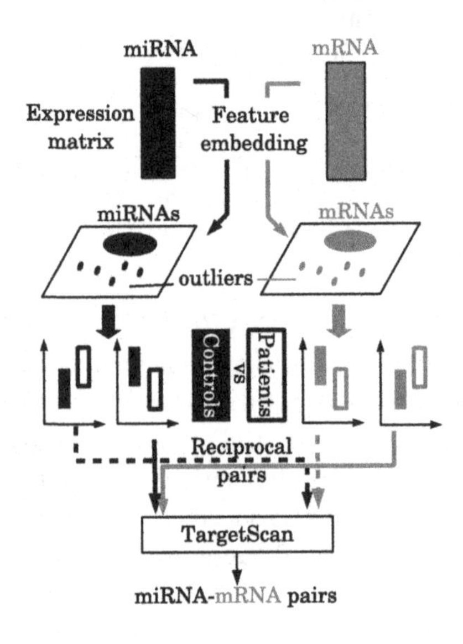

Fig. 8.1 Schematic of miRNA-mRNA interactions using PCA based unsupervised FE.

Tables 8.2 and 8.3 shows the summary of pre-screened miRNAs and mRNAs used and summary of identified miRNA-mRNA interactions, respectively (for more details including the list of pairs identified, see the original study [Taguchi (2016a)]). As can be seen, in spite of that we used single same adjusted P-values threshold, the numbers of pre-screened miRNAs and mRNAs do not drastically vary even when the number of samples used varies. Although samples are not matched and measurements were performed with diverse platforms (microarrays), we could successfully identify miRNA-mRNA pairs for all cancers investigated.

In this demonstration, one may be able to understand that not samples but features (mRNAs and miRNAs) based P-value estimation is very useful and suitable for biological researches.

Table 8.2 Summary of identification of miRNAs and mRNAs screened by PCA based unsupervised FE.

		Sample numbers		Probe numbers	
Cancers	GEO ID	Patients	Normal tissue	Selected	Not selected
HCC					
mRNA	GSE45114	24	25	269	22963
miRNA	GSE36915	68	21	58	1087
NSCLC					
mRNA	GSE18842	46	45	1098	53504
miRNA	GSE15008	187	174	268	3428
ESCC					
mRNA	GSE38129	30	30	189	22088
miRNA	GSE19337	76	76	37	1217
PC					
mRNA	GSE21032	150	29	399	43020
miRNA	GSE84318	27	27	23	700
CRC/CC					
mRNA	GSE41258	186	54	309	21974
miRNA	GSE48267	30	30	12	839
BC					
mRNA	GSE29174	110	11	980	33600
miRNA	GSE28884	173	16	18	2258

Selected/not selected mean if mRNAs and miRNAs are selected by PCA based unsupervised FE. GEO ID represents ID in Gene expression omnibus [GEO (2016)].

Table 8.3 Identification of mRNAs targeted by miRNAs, using TargetScan. Pairs shows the number of miRNA-mRNA pairs included in TargetScan within miRNAs and miRNAs selected in Table 8.2. Numbers of miRNA and mRNAs are those within pairs.

Cancers	Pairs		miRNA		mRNA	
HCC	20	(9)	13	(13)	18	(16)
NSCLC	311	(184)	27	(27)	113	(72)
ESCC	4	(2)	3	(3)	4	(4)
PC	32	(18)	8	(8)	19	(6)
CRC/CC	8	(3)	7	(7)	7	(6)
BC	37	(17)	11	(11)	30	(25)

The number of pairs are less than those of mRNAs and/or miRNAs, since multiple pairs share the same mRNAs and miRNAs. The numbers in parenthesis: for mRNAs and miRNAs, those associated previous studies that papers relation with cancers are counted. For pairs, those associated with negative correlation between miRNAs and mRNAs in starbase [Li *et al.* (2014)] were counted.

8.4.2 *miRNA-mRNA interaction in PTSD mediated heart disease*

In the previous subsection, we successfully demonstrated that PCA based unsupervised FE can be applicable to pre-screen mRNAs and miRNAs

Table 8.4 Samples of stressed conditions.

	1	2	3	5	10
1	T	T,C	T	T,C	T,C
10	-	-	-	T,C	-
42	-	-	-	-	T,C

Row: rest days, column: caged days. T: treated (stressed), C: control. Each conditions are associated with four replicates.

associated with distinct expression between tumors and normal tissues. However, the true usefulness of this methodology can be enhanced when dealing with samples in multiple classes.

In this subsection, we demonstrate the usefulness of PCA based unsupervised FE in categorical multiclass problem, by considering post-traumatic stress diseases (PTSD) mediated heart failure. PTSD often takes place when human beings face the life-threading stresses, e.g., serving as a solder in battle field. PTSD was also widely observed in 9/11 attack in USA [Gargano *et al.* (2015)]. Although PTSD primarily causes mental problems, PTSD also affects. Among those affected organs, heart often exhibits disorder. Since heart beats are dependent upon mental conditions, it is not surprising for PTSD to cause heart problems. However, how PTSD mediates heart failure was not known well.

miRNAs were also considered to be one of candidates that mediate heart failure. Recently, Cho et al [Cho *et al.* (2014)] measured miRNA and miRNAs expression in the stressed mice hearts. From the point of data analysis strategy, multiple problems are associated with these data sets. First, it is better to identify coincidence between mRNA and mRNAs expression, since such kind of coincidence can be an evidence of more possible functional pairs of mRNA and miRNAs. This will enable us to understand the mechanisms of PTSD mediated heart disease. Second, since the experimental conditions cannot be simply ranked, how to identify differential expression between treated and control samples is not obvious. Both difficulties can be resolved using PCA based unsupervised FE [Taguchi *et al.* (2015b)].

Detail of Cho *et al.* [Cho *et al.* (2014)]'s experiments is as followed. Mice are caged with violent mice during specified period (X days) and heart gene expression was measured Y days after the isolation from violent mice. Controls are treated samely excluding being caged without violent mice. Since multiple combinations of periods X and Y days are tried, this is categorical multiclass problem. We have applied PCA based unsupervised FE to mRNAs and miRNAs expression separately with combining all X and Y days combinations (Table 8.4). Since we do not need any information about ranking, this procedure resolve the second problems mentioned in the above.

We first identified 100 outlier miRNAs using PCA based unsupervised FE [Taguchi *et al.* (2015b)]. Then in order to see if PCA based unsupervised FE successfully identified miRNAs differentially expressed between stressed and control mice, t test was applied to top most 100 miRNAs outliers selected by PCA based unsupervised FE. Although PCA based unsupervised FE did not use category information to select mRNAs at all, selected 100 miRNAs turned out to be expressed differently and significantly between all five pairs of stressed and control mice heart. This shows the ability of PCA based unsupervised FE to identify critical miRNAs even without using category information. The same procedure was repeated to 100 mRNAs outliers selected by PCA based unsupervised FE as well. We found that PCA based unsupervised FE successfully identified miRNAs differently and significantly expressed between all five pairs of stressed and control mice hearts as well. Thus, again PCA based unsupervised FE can identify critical mRNAs even without category information.

One may wonder why PCA based unsupervised FE could simultaneously identify miRNAs or mRNAs expressed differently between multiple pairs of treated and control samples without using the information to which category samples belong; it looks like discrepancy. In principle, what PCA based unsupervised FE detect is not a ranked order of differential expression but a group behavior of genes. And group behavior is likely reflecting distinction between treated and control samples, since it should be the most significant distinction among analyzed samples. This is the reason why PCA based unsupervised FE can identify mRNAs and miRNAs that exhibit differential expression between treated and control samples.

Next, we tried to investigate the coincidence between mRNAs and miRNAs. First, we checked if loading, $\boldsymbol{v}_k$, of PC whose scores are used for outlier identification are negatively corrected between mRNAs and miRNAs. Figure 8.2 shows the (averaged) scatter plots of loadings of the first principal component (PC1) used for outliers identification. P values attributed to Pearson correlation coefficients are 0.01. Since the correlation coefficient increased after averaging within experimental conditions, negative correlation between gene expression and promoter methylation is significantly associated with experimental conditions. Second, we computed the Pearson correlation coefficients of raw mRNA and miRNA expression for 47 pairs of miRNAs and miRNAs, between which seed macth was expected. Then we found that 45 out of 47 pairs are associated with negative correlation. Since we did not assume the negative correlation when identifying

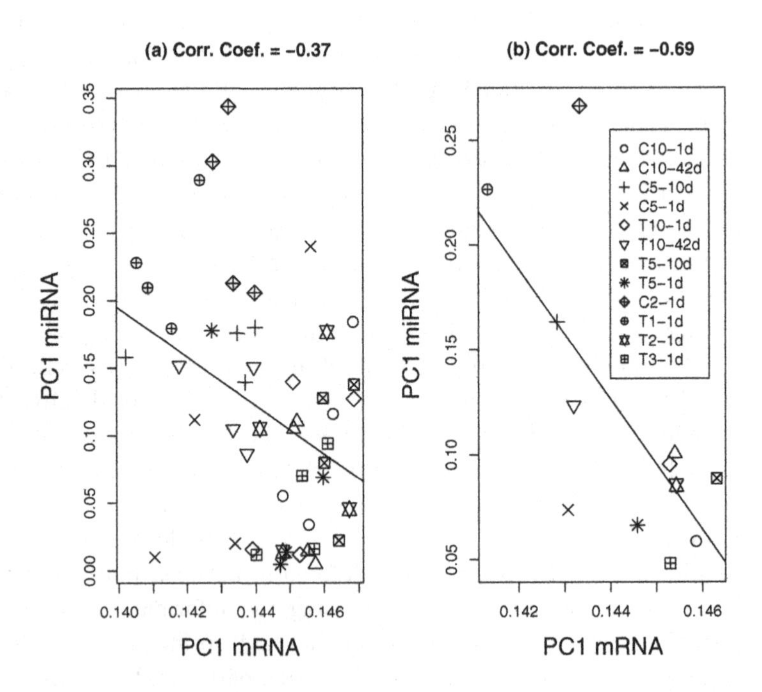

Fig. 8.2 Left: Scatter plot of PC1 loading, $\boldsymbol{v}_1$, between gene expression (vertical axis) and gene expression (horizontal). P-values attributed to correlation coefficient is 0.01. Right: Those averaged with experimental condition. P-values attributed to correlation coefficient is 0.01. T/CX-Yd stands for treated (T) or control (C) samples for Y days' rest after X days spent in the cage with/without violent mice.

outliers, the high ratio of negative correlation suggests that superiority of our methodology, since no other methods tested could not achieve such a high ratio [Taguchi *et al.* (2015b)].

Although we have further confirmed the biological feasibilities of selected mRNAs and miRNAs using various annotation servers, we ommit the part. Please see the original paper [Taguchi *et al.* (2015b)] for more details.

In conclusion, PCA based unsupervised FE was successfully applied to miRNA and mRNA interactions in various tumors as well as PTSD mediated heart diseases. This methodology is useful to identify miRNA-mRNA interaction in wider range of biological problems.

8.5 Integrated analysis of gene expression and promoter methylation

In the previous subsection, we demonstrated that PCA based unsupervised FE is useful to integrate the analysis of mRNA and miRNA expression. Since miRNA-mRNA interaction is many-to-many, the selected miRNAs and mRNAs have more opportunities to be paired. Thus, next question is if PCA based unsupervised FE can identify pairs even within one-to-one interaction. It should be much harder since each one has only one (or very limited number of) possible partner to be paired.

As an example for identification of one-to-one interaction, we consider interaction between promoter methylation and mRNA expression. Although it is generally supposed for hypermethylation to suppress gene expression, it is also known to be highly context dependent [Moarii *et al.* (2015)]. Thus, it is definitely important to identify which pairs of promoter methylation and gene expression are related in a specific sample. In the following subsections, we report the trials that use PCA based unsupervised FE to this task.

8.5.1 *Target genes of epigenetic therapy of non-small cell lung cancer*

In spite of numerous proposals of various therapy, non-small cell lung cancer still remains lethal [Counago *et al.* (2016)]. Among those newly proposed, epigenetic therapy is regarded to be a promising candidate. Although there are numerous reports reporting efficiency of epigentic therapy towards non-small cell lung cancer [Forde *et al.* (2014)], no genetic backgrounds were known. In this regard, integrated analysis of gene expression and promoter methylation is helpful. Since no known established *in vitro* experiments that can measure the effect of epigenetic therapy towards non-small cell lung cancer exist, we instead investigated gene expression and promoter methylation of reprogrammed non-small cell lung cancer cell line. Both reprogramming and epigenetic therapy are expected to alter epigenome. Thus, we can expect that these two share the gene expression alternation.

In order to identify genes associated with aberrant gene expression and aberrant promoter methylation during reprogramming, data set obtained by Mahalingam et al [Mahalingam *et al.* (2012)] were analyzed by PCA based unsupervised FE [Taguchi *et al.* (2016)]. Table 8.5 shows the samples used for their analysis.

Table 8.5 Samples used in this study. Numebers are those of biological replicates.

Cell lines	H1	H358	H460	IMR90	iPCH358	iPCH460	iPSIMR90	piPCH358
mRNA	3	3	3	3	3	3	3	3
Methylation	3	3	3	3	3	3	3	3

H1: ES cell, H358,H460: NSCLC, IMR90:Human Caucasian fetal lung fibroblast, iPCH358, iPCH460, iPSIMR90: reprogrammed cell lines, piPCH358:re-differentiated iPCH358

Gene embedding was performed separately towards mRNA expression and promoter methylation. 24 PC loadings were obtained. In order to identify PCs whose loading are highly correlated between mRNA expression and prompter methylation, we have applied hierarchical clustering PC loading (Fig. 8.3). Eventually, PC loadings are the most strongly corre-

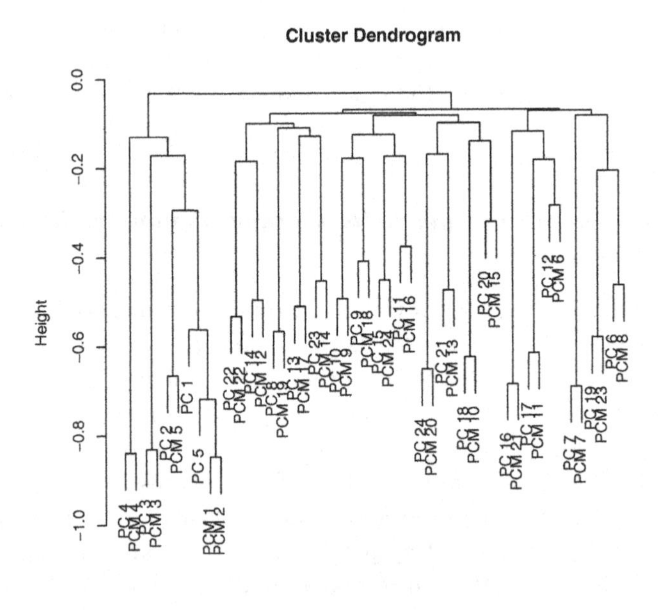

Fig. 8.3 Hierarchical clustering of 24 PC loadings obtained using mRNAs (PC) and promoter methylation (PCM), respectively. Distances are negative signed absolute Pearson correlations.

lated for the third and fourth PCs between mRNA expression and promoter methylation. Thus, we decided to identify outliers using the third

Table 8.6 List of genes identified by PCA based unsupervised FE in non-small cell lung cancer cell line reprogramming.

PC3	F2R DKK3 SFRP1 SLC16A12 HOXA5 KIF1A H2AFY ATP5G2 TM4SF1 GPR56 S100P
PC4	SPINT2 CDH1 LAMC2 HMGA1 LAD1 PFKFB3 DEFB1 SRGN UCHL1 ALDH3A1 EPB41L3 RTN1 LAMA1

and the fourth PC scores. Top most 300 outliers are identified using the third and fourth PC scores of mRNA expression and promoter methylation, respectively. Genes selected commonly between mRNA expression and promoter methylation via either the third or fourth PC scores are listed (Table 8.6). Although we cannot detail the biological meanings of PC3 and PC4 here because of the lack of spaces, the outline is as follows. Both PC3 and PC4 represent distinction between pre and post reprogrammed non-small cell lung cancer cell lines. In addition to this, PC3 and PC4 also shows that the coincidence between reprogrammed non-small cell lung cancer cell lines and pluripotent/iPS cell lines. Furthermore, PC3 and PC4 confirmed that pluripotent/iPS cell lines are distinct from IMR90 that is not reprogrammed cell line. One may also wonder why there are two PCs obtained. The distinction between two PCs are coincidence between two non-small cell lung cancer cell lines. PC3 represent aberrant but coincident mRNA expression/promoter methylation between non-small cell lung cancer and reprogramed one while PC4 represent aberrant but opposite mRNA expression/promoter methylation between non-small cell lung cancer and reprogramed one. Anyway, it is obvious that PCA based unsupervised FE can have superior power to identify biologically meaningful features even in categtorical multiclass problems in an unsupervised manner. To our knowledge, no other methods can do this. For more details, see original paper [Taguchi *et al.* (2016)].

Among those selected (Table 8.6), due to massive literature search, we identified that SFRP1 was the most promising candidate as epigenetic therapy target gene in non-small cell lung cancer because of the following two reasons. First, SFRP1 was highly expressive in histone deacethylation inhibitor (HDACi) non-resistant non-small cell lung cancer cell lines than on HDACi resistant non-small cell lung cancer cell lines [Miyanaga *et al.* (2008)]. Second, histone acetylation of SFRP1 was enhanced due to HDACi [Tang *et al.* (2010)]. These two strongly suggested that SFRP1 was a promising candidate of epigenetic therapy. Although we have done more researches on this topics, including GO term enrichment analysis and

Table 8.7 The number of samples of each cell lines used.

Cell lines	HTB56		A549	
	With	Without	With	Without
	metastasis			
mRNA expression	3	3	3	3
Promoter methylation	2	2	2	2

protein-protein docking simulation between SFRP1 and its target protein, more details are available in the original paper [Taguchi *et al.* (2016)].

8.5.2 *Metastasis causing genes of non-small cell lung cancer*

The second example of integrated analysis of mRNA expression and promoter methylation is also non-small cell lung cancer, but in distinct concept: identification of metastasis causing genes. As mentioned in the above, non-small cell lung cancer is lethal. Especially after metastasis, more than half will die within one year. Thus, suppressing metastasis in non-small cell lung cancer is critically important. In this purpose, we re-analyzed two non-small cell lung cancer cell lines' mRNA expression and promoter methylation pre and post metastasis [Hascher *et al.* (2014)] using PCA based unsupervised FE [Umeyama *et al.* (2014)].

Table 8.7 shows the number of samples used in this study. As usual, PCA based unsupervised FE was applied to mRNA expression and promoter methylation, separately. In contrast to the integrated analyses of mRNA expression promoter methylation in the previous subsection, PC loading behaves in a little bit more complex manner. The first PC loading exhibits no sample dependence. Although the second PC loading exhibit some sample dependence, it is not between pre and post metastasis, but between two cell lines. The third PC exhibits distinction between pre and post metastasis only for HTB56 cell lines. Distinction between pre and post metastasis in A549 cell lines appear in PC5 (mRNA expression) and PC4 (promoter methylation). This suggests that difference between pre and post metastasis is very little and can be hardly identified.

Fortunately, we could identified some genes commonly between top most 100 outliers using mRNA expression/promoter methylation PC3 vs PC3 or PC5 vs PC4 scores. Table 8.8 shows the list of genes selected. One may wonder if these numbers are small enough to obtain accidentally. However, since total number of mRNAs and possible methylation cites are more than 10^4, P-values that these overlap occurs accidentally are 3.5×10^{-5} for PC3

Table 8.8 Genes selected by PCA based unsupervised FE during metastasis development.

mRNA expression	Promoter methylation	Genes
PC3	PC3	HOXB2, CCDC8, ZNF114, DIO2, LAPTM5, RGS1, B3GALNT1
PC5	PC4	TINAGL1, PMEPA1, CX3CL, ICAM1

vs PC3 and 5.1×10^{-4} for PC5 vs PC4, respectively. Thus, these overlaps cannot be accidental and these genes are worthwhile considering.

Although we have performed extensive researches including pathway/GO term enrichment analysis as well as *in silico* drug discovery [Umeyama *et al.* (2014)], we cannot discuss about it because of lack of spaces. Finally, we identify two promising metastasis causing genes, TINAGL1 and B3GALNT1. Although there were no experimental studies that support our findings, after the publication of our study [Umeyama *et al.* (2014)], Takahashi *et al.* [Takahashi *et al.* (2016)] reported that TINAGL1 plays potential roles in mouse impaired female fertility that is supposed to be related to metastasis. Thus, our findings may be feasible.

In conclusion, even if the distinction is very little and the number of samples is small, PCA based unsupervised FE can identify critical genes in an unsupervised manner.

8.5.3 *Genes mediating transgenerational epigenetics*

The third example of integrated analysis of mRNA expression and promoter methylation is transgenerational epigenetics [Nagy and Turecki (2015)]. Usually, phenotypes obtained after the birth, i.e., acquired traits, are supposed not to be inherited. However, a part of epigenome is believed to be inherited and can be of course altered even after the birth. The concept of transgenerational epigenetics refers to heritages of acquired traits through epigenome. Although it is not easy to investigate experimentally transgenerational epigenetics, recently rodent model turned out to be useful to investigate transgenerational epigenetics. Especially, the effects of endocrine disrupting compounds were extensively studied [Rissman and Adli (2014)]. Although numerous reports suggested that endocrine disruptors cause multiple diseases through transgenerational epigenetics, no known mechanisms exist yet. In this subsection, in order to understand how endocrine disruptors mediate transgenerational epigenetics, we re-analized aberrant mRNA

Table 8.9 Samples of primordial germ cells between E13 and E16 rat F3 generation vinclozolin lineage. Treated means vinclozolin treatments. Promoter methylation was given as ratio between control and treated samples.

Time	E13		E16	
	Treated	Control	Treated	Control
mRNA expression	2	2	2	2
Promoter methylation	3		3	

expression and promoter methylation in primordial germ cells between E13 and E16 rat F3 generation vinclozolin lineage by PCA based unsupervised FE [Taguchi (2015)]. Table 8.9 shows the list of samples used for this study [Skinner *et al.* (2013)]. As usual, PCA based unsupervised FE was applied to mRNA expression and promoter methylation separately. Then the second PC loading of mRNA expression and the third PC loadings for promoter methylation were identified to exhibit differential expression between E13 and E16. In this case, we need to increase the number of genes selected for mRNA expression or promoter methylation to 1,000 in order to get significant overlaps. Then 48 genes (not shown here because of too many numbers. The list of genes is available in the original paper [Taguchi (2015)]) are identified as commonly selected genes.

Although one may wonder that this achievement is as good as usual, since we had to identify as many as 1000 genes associated with aberrant mRNA expression or promoter methylation, it is a very difficult problem. Actually, Skinner et al [Skinner *et al.* (2013)] who have done the original analysis clearly denoted that they could not identify genes simultaneously associated with aberrant mRNA expression and aberrant promoter methylation. Promoter methylation must be reset every time development starts in new generation. This suggested that no promoter methylation can be inherited as it is. Thus if something is inherited, only possibility is timing of promoter methylation. This is the reason why Skinner et al tried to compare mRNA expression as well as promoter methylation at the distinct time points, E13 and E16. Although we also tried to identify genes simultaneously associated with aberrant mRNA expression and aberrant promoter methylation using other methodologies including limma as well as SAM than PCA based unsupervised FE, biological significance of selected genes was substantially inferior to those identified PCA based unsupervised FE.

Since we cannot discuss more details about this study because of lack of spaces, please refer to the original study [Taguchi (2015)] for more details.

Table 8.10 Samples of HCC and CCC.

	HCC	CCC	Adjusted tissue
mRNA	10	6	16
miRNA	10	6	16
compounds	10	6	16

8.6 Integration of more than two multi-omics data sets

Although in the previous sections, we demonstrated that PCA based unsupervised FE can integrate two omics: mRNA and miRNA profiles as well as mRNA expression and promoter methylation. However, there are other omics data available than those three. If more than two omics layers are simultaneously integrated, more feasible interpretation are expected.

Recently, we demonstrated that PCA based unsupervised FE can integrated three distinct omics layers without changing strategy described in the above much [Murakami *et al.* (2015)]. The purpose of our research is to identify distinction between hepatocellular carcinoma (HCC) and cholangiocarcinoma (CCC). These two cancers are both liver cancers, but the latter is more difficult to treat with. Thus, primarily, it is important to discriminate between two cancers. However, at the moment how these two cancers develop is not well known.

Table 8.10 shows the data set used in this study. Ten HCC and six CCC patients' liver with normal (adjusted) organs were also extracted and mRNA, miRNA and compounds are simultaneously measured (i.e., matched samples). PCA based unsupervised FE was applied to mRNA, miRNA and compounds separately as usual. Since it is a categorical multi-classes (four classes) problem, we performed hierarchical clustering again (Fig. 8.4). Since we would like to find PC loadings attributed to samples correlated within miRNA, mRNA and compounds, we decided to employ the first and the second PCs for mRNA and miRNA, the third PC for compounds. Figure 8.5 shows the scatter plots as well as correlation coefficients of these PC loadings. It is obvious that each PC loadings is significantly correlated with more than one PC loadings other than itself. Then compounds, mRNAs and miRNAs that are top ranked outliers along corresponding PC scores are selected. Since this study was done substantially long time ago before the procedure introduced in this manuscript was established, the number of outliers varies from omics to omics. For more details see the original paper [Murakami *et al.* (2015)].

Since the primary purpose is to identify potential biomarkers, we tried to discriminate HCC, CCC and normal tissues using identified mRNA, miRNA

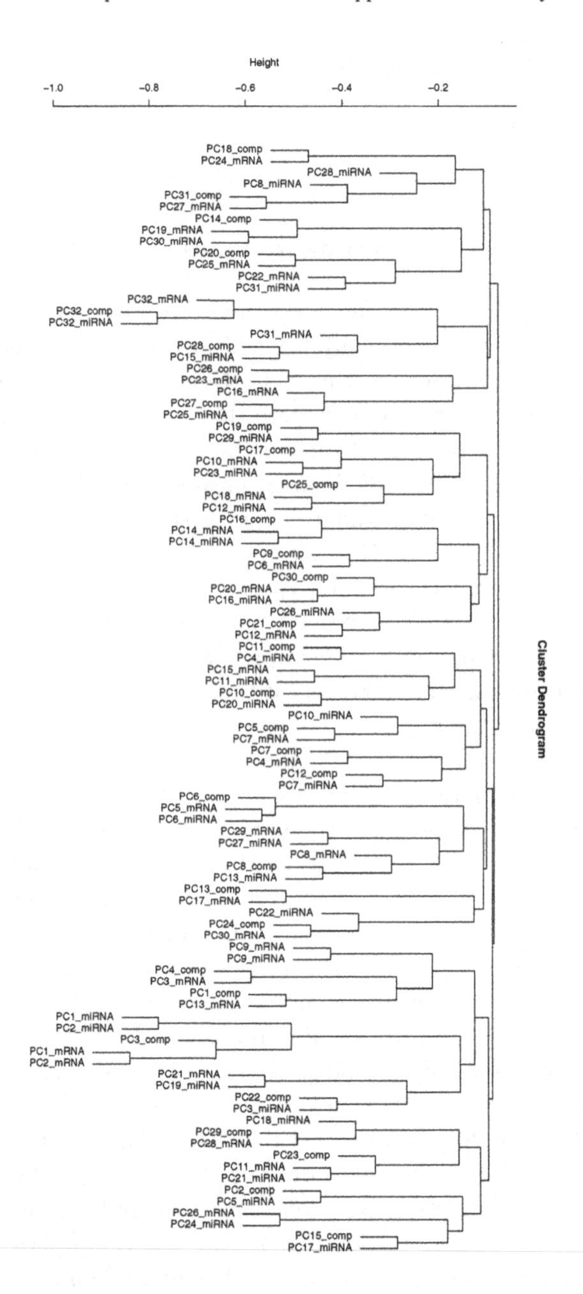

Fig. 8.4 Hierarchical clustering of 16 PC loadings for mRNA, miRNA and compounds. Distance is negative signed absolute Pearson correlation coefficients.

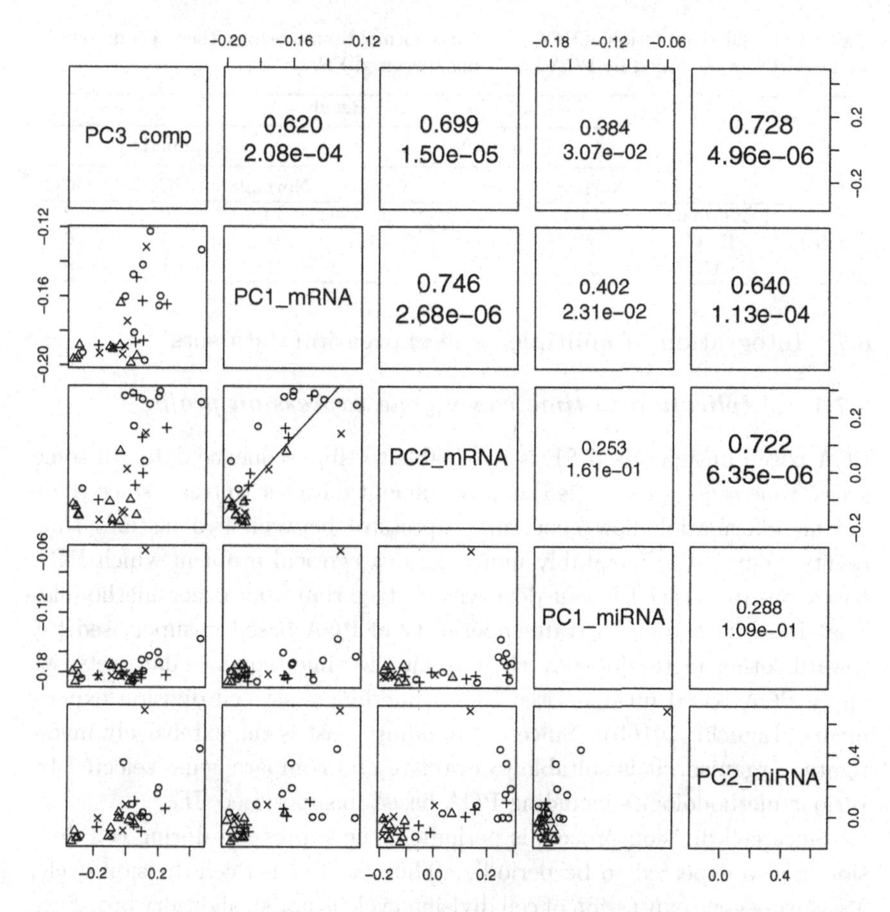

Fig. 8.5 Left lower triangle:Scatter plots of PC loadings used for outliers identification. ○ :CCC, △: adjusted normal tissue for CCC, +:HCC, ×: adjusted normal tissues for CCC. Right upper triangle: Pearson correlation coefficients as well as attributed *P*-values.

and compounds (Table 8.11). Although discrimination using mRNA fails, miRNA or compounds identified successfully discriminate HCC, CCC and normal tissue (accuracy is 84 % and 78 %, respectively). Thus, PCA based unsupervised FE successfully identified, in an unsupervised manner, outliers that can discriminate between HCC, CCC and normal tissues well. Although we have further discussed biological consequences of these selected mRNAs, miRNAs and compounds, we omit these because of lack of spaces. Please refer the original paper [Murakami *et al.* (2015)] for more details.

Table 8.11 Discrimination of HCC, CCC and normal tissues using either 14 compounds or 17 miRNAs identified by PCA based unsupervised FE.

		Result					
		miRNA			Compounds		
		Normal	HCC	CCC	Normal	HCC	CCC
	Normal	13	1	1	14	0	2
Predict	HCC	2	4	1	0	5	0
	CCC	1	1	8	2	1	8

8.7 Integration of multiple gene expression data sets

8.7.1 *Application to time cause gene expression profile*

PCA based unsupervised FE is applicable to time sequence data. In some sense, time sequence data is the most difficult data set to treat, since there are no information how genes are expressive between two distinct time points. Thus, it is inevitably multi-class categorical problem which PCA based unsupervised FE can deal with better than any other methodologies. In order to demonstrate superiority of PCA based unsupervised FE towards other methodologies when applied to time sequence data sets, we apply PCA based unsupervised FE to budding yeast cell division experiments [Taguchi (2016b)]. Since the budding yeast is the extensively investigated organism, it is suitable to evaluate and compare genes selected by various methodologies including PCA based unsupervised FE.

Since cell division process is periodic, gene expression during cell division is also expected to be periodic. Thus, it is called cell division cycle. Measuring gene expression of cell division cycle is not straightforward, since individual cell (e.g., budding yeast) usually does not cause cell division synchronically, we need to force yeast to be synchronized.

In this study, we consider two independent methodologies of synchronization. The first one is to control feeding. Without feeding, cell division process is terminated, thus it is possible to arrest cell division (This is called yeast metabolic cycle, YMC). The second one is more direct way; using *cdc* mutant. Making use of temperature sensitivity, budding yeast can be arrested in fixed phase of cell division cycle (This is called yeast cell division cycle, YCDC).

As for the first example, we employed Tu *et al.* [Tu *et al.* (2005)]'s YMC experiments. In this study, they observed 36 time points that are supposed to correspond to a length of three YMCs. After gene embedding applied, the scatter plots for the first four PC loading as well as corresponding

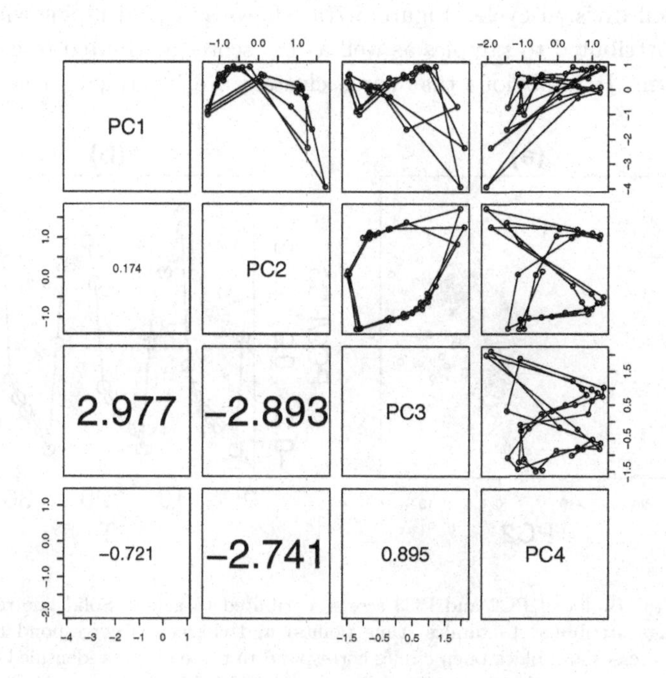

Fig. 8.6 Upper right triangle: Scatter plots of the first four PC loadings obtained by gene embedding. Adjusted time points are connected by solid lines. Lower left triangle: winding numbers computed from the corresponding scatter plots.

widing numbers was drawn (Fig. 8.6) in order to identify PC loadings used for FE. Here winding number was the measure how many times time points rotate around center of mass in the two dimensional embedding space. As can be seen in Fig. 8.6, maximum winding number is three, which is equivalent to the expected period of three. Thus, it is obvious that PCA based unsupervised FE has ability to deal with periodic time sequence data, in an unsupervised manner, without specifying time period.

In addition to this, Fig. 8.6 clearly identifies period doubling (i.e., eight character shaped trajectories). It definitely shows the superiority of PCA based unsupervised FE that can identify period doubling that no other supervised methodology can identify, since supervised methodologies seek only periods coincident with cell division cycle.

In order to further confirm the effectiveness, we also compare our results with the previous study [Tu *et al.* (2005)]. Tu et al [Tu *et al.* (2005)] identify three distinct functional clusters of genes associated with three peaks during cell division cycle. Figure 8.7(a) shows so called biplot where PC loading attributed to samples as well as PC scores attributed to genes are overdrawn. It is obvious that three clusters are identified. Due to GO

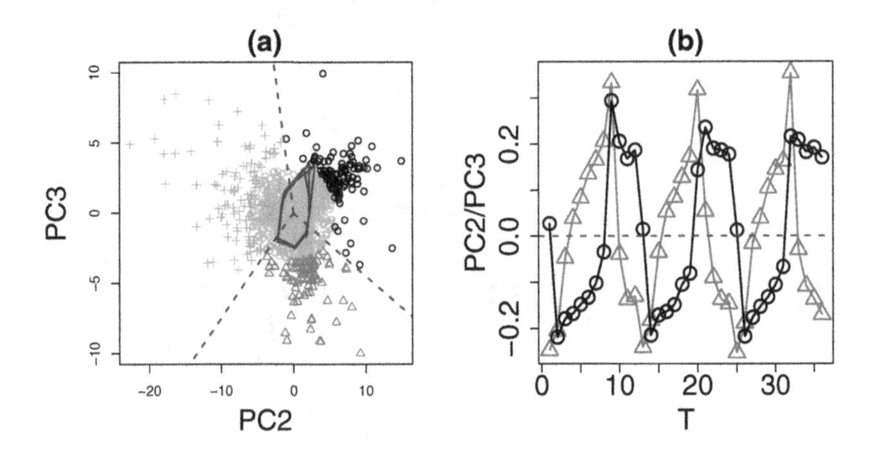

Fig. 8.7 (a) Biplot of PC2 and PC3 scores attributed to genes. Solid line represents PC loadings attributed to samles (time points) and characters correspond to genes. Triangles, crosses and black open circles correspond to three clusters identified applying K-means to gene identified by PCA based unsupervised FE within this embedding space (gray open circles correspond to genes not identified by PCA based unsupervised FE). Broken lines emphasize that circular variable can separate three clusters. (b) Time dependence of PC2($\circ$)/PC3($\triangle$) loadings.

term/KEGG pathway enrichment analysis, these three clusters are associated with biological functions consistent with those Tu *et al.* identified (for details, see original paper [Taguchi (2016b)]). In addition to this, periodic forms of time dependence heavily deviates from sinusoidal function (Fig. 8.7(b)). This suggested that sinusoidal fitting that was often employed can be erroneous. All of these suggested the superiority of unsupervised methodology that does not assume period length as well as sinusoidal functional form.

As for the second example (YCDC), we applied PCA based unsupervised FE to seven budding yeast gene expression out of eight stored in CycleBase [Santos *et al.* (2015)]. Although we cannot detail the results obtained because of lack of space, PCA based unsupervised FE was successfully applied to these seven gene expression profiles. For example, among separately identified outlier genes within each profile, there are 37 genes commonly selected in more than or equal to six out of seven profiles. Since PCA based unsupervised FE select less than 200 genes among more than thousands genes for each profile, this coincidence is highly significant. Enrichment analysis performed also supports the feasibility of 37 selected genes. This suggested that PCA based unsupervised FE has ability of integrating as many as seven gene expression profiles in an unsupervised manner. For more details, see the original paper [Taguchi (2016b)].

8.7.2 *Application to amyotrophic lateral sclerosis*

Amyotrophic lateral sclerosis (ALS) is a difficult disease to treat because of multiple reasons. First, we are not sure what causes ALS. Although there are some evidences that suggest the relationship between ALS and genetic background, no known genetic mutation that is firmly related to ALS. Second, since ALS is the disease of motor neuron, it is not easy to get clinical samples. If one removes motor neuron from patients, this treatment itself injures patients. This is similar to heart or brain diseases. Third, there are limited model animals for this diseases. As a results, both *in vitro* and *in vivo* study of this disease are not easy.

There are limited studies to investigate gene expression analysis of ALS because of the reasons mentioned in the above. Recently, we have analyzed [Taguchi *et al.* (2015a)] fibroblast cell line gene expression taken from patients as well as healthy controls [Fogel *et al.* (2014)].

Fogel et al [Fogel *et al.* (2014)] generated fibroblast cell line from patients and measured gene expression between healthy control and patients. In addition to this, cell lines were transfected by three mutated genes that are known to cause symptom. Table 8.12 shows the list of samples used in this study. We have integrated these gene expression profiles as follows. First,

Table 8.12 Fibroblast cell lines gene expression profiles used for this study.

	Transfection				
	Nothing	Mock	Mutation1	Mutation2	Mutation 3
Patients	2	2	2	2	2
Healthy controls	2	2	2	2	2

four expression profiles without transfection and remaining 16 gene expression profiles were analyzed separately. The former identified 708 genes while the latter identified 715 genes as outliers associated with adjusted P-values less than 0.01 by PCA based unsupervised FE. Interestingly they are highly coincident with each other. For example, more than half of genes (393 genes) are common. KEGG pathway enrichment analyses were also almost identical between genes identified in 4 samples without transfection and those in 16 samples with transfections. These suggested that PCA based unsupervised FE can integrate multiple gene expression as well. As mentioned in the above, although it is motor neuron disease, slight difference observed in fibroblast gene expression can be detected correctly by our methodology. Thus, PCA based unsupervised FE also has ability to detect very slight difference in gene expression.

8.8 Conclusions

In this chapter, we have introduced PCA based unsupervised FE. It turned out to successfully integrate multiple omics profiles, e.g.,

- mRNA expression and promoter methylation
- mRNA and miRNA expression,
- mRNA, miRNA expression and metabolome (compounds).

It could also be applied to times sequence data exhibiting periodicity. Finally, we also demonstrated that it can integrated multiple gene expression. All of them can be regarded to be multi-class categorical data set. Since no other method can deal with multi-class categorical data set well, our methodology is promising.

References

Agarwal, V., Bell, G. W., Nam, J. W., and Bartel, D. P. (2015). Predicting effective microRNA target sites in mammalian mRNAs, *Elife* **4**.

Artmann, S., Jung, K., Bleckmann, A., and Beissbarth, T. (2012). Detection of simultaneous group effects in microRNA expression and related target gene sets, *PLoS ONE* **7**, 6, p. e38365.

Benjamini, Y. and Hochberg, Y. (1995). Controlling the false discovery rate: A practical and powerful approach to multiple testing, *Journal of the Royal Statistical Society. Series B (Methodological)* **57**, 1, pp. 289–300, `http://www.jstor.org/stable/2346101`.

Bleckmann, A., Leha, A., Artmann, S., Menck, K., Salinas-Riester, G., Binder, C., Pukrop, T., Beissbarth, T., and Klemm, F. (2015). Integrated miRNA and mRNA profiling of tumor-educated macrophages identifies prognostic subgroups in estrogen receptor-positive breast cancer, *Mol Oncol* **9**, 1, pp. 155–166.

Cho, J. H., Lee, I., Hammamieh, R., Wang, K., Baxter, D., Scherler, K., Etheridge, A., Kulchenko, A., Gautam, A., Muhie, S., Chakraborty, N., Galas, D. J., Jett, M., and Hood, L. (2014). Molecular evidence of stress-induced acute heart injury in a mouse model simulating posttraumatic stress disorder, *Proc. Natl. Acad. Sci. U.S.A.* **111**, 8, pp. 3188–3193.

Counago, F., Rodriguez, A., Calvo, P., Luna, J., Monroy, J. L., Taboada, B., Diaz, V., and Rodriguez de Dios, N. (2016). Targeted therapy combined with radiotherapy in non-small-cell lung cancer: a review of the Oncologic Group for the Study of Lung Cancer (Spanish Radiation Oncology Society), *Clin Transl Oncol* .

Ding, M., Li, J., Yu, Y., Liu, H., Yan, Z., Wang, J., and Qian, Q. (2015). Integrated analysis of miRNA, gene, and pathway regulatory networks in hepatic cancer stem cells, *J Transl Med* **13**, p. 259.

Fogel, B. L., Cho, E., Wahnich, A., Gao, F., Becherel, O. J., Wang, X., Fike, F., Chen, L., Criscuolo, C., De Michele, G., Filla, A., Collins, A., Hahn, A. F., Gatti, R. A., Konopka, G., Perlman, S., Lavin, M. F., Geschwind, D. H., and Coppola, G. (2014). Mutation of senataxin alters disease-specific transcriptional networks in patients with ataxia with oculomotor apraxia type 2, *Hum. Mol. Genet.* **23**, 18, pp. 4758–4769.

Forde, P. M., Brahmer, J. R., and Kelly, R. J. (2014). New strategies in lung cancer: epigenetic therapy for non-small cell lung cancer, *Clin. Cancer Res.* **20**, 9, pp. 2244–2248.

Fu, J., Tang, W., Du, P., Wang, G., Chen, W., Li, J., Zhu, Y., Gao, J., and Cui, L. (2012). Identifying microRNA-mRNA regulatory network in colorectal cancer by a combination of expression profile and bioinformatics analysis, *BMC Syst Biol* **6**, p. 68.

Gargano, L. M., Caramanica, K., Sisco, S., Brackbill, R. M., and Stellman, S. D. (2015). Exposure to the World Trade Center Disaster and 9/11-related post-traumatic stress disorder and household disaster preparedness, *Disaster Med Public Health Prep* **9**, 6, pp. 625–633.

GEO (2016). Gene expression omnibus, http://www.ncbi.nlm.nih.gov/geo/.

Hascher, A., Haase, A. K., Hebestreit, K., Rohde, C., Klein, H. U., Rius, M., Jungen, D., Witten, A., Stoll, M., Schulze, I., Ogawa, S., Wiewrodt, R., Tickenbrock, L., Berdel, W. E., Dugas, M., Thoennissen, N. H., and Muller-Tidow, C. (2014). DNA methyltransferase inhibition reverses epigenetically embedded phenotypes in lung cancer preferentially affecting polycomb target genes, *Clin. Cancer Res.* **20**, 4, pp. 814–826.

Jansson, M. D. and Lund, A. H. (2012). Microrna and cancer, *Molecular Oncology* **6**, 6, pp. 590–610, doi:http://dx.doi.org/10.1016/j.molonc.2012.09.006, `http://www.sciencedirect.com/science/article/pii/S1574789112000981`, cancer epigenetics.

Jonnalagadda, S. and Srinivasan, R. (2008). Principal components analysis based methodology to identify differentially expressed genes in time-course microarray data, *BMC Bioinformatics* **9**, p. 267.

Krzanowski, W. J. (1987). Selection of variables to preserve multivariate data structure, using principal components, *Journal of the Royal Statistical Society. Series C (Applied Statistics)* **36**, 1, pp. 22–33, `http://www.jstor.org/stable/2347842`.

Li, J. H., Liu, S., Zhou, H., Qu, L. H., and Yang, J. H. (2014). starBase v2.0: decoding miRNA-ceRNA, miRNA-ncRNA and protein-RNA interaction networks from large-scale CLIP-Seq data, *Nucleic Acids Res.* **42**, Database issue, pp. D92–97.

Li, X., Gill, R., Cooper, N. G., Yoo, J. K., and Datta, S. (2011). Modeling microRNA-mRNA interactions using PLS regression in human colon cancer, *BMC Med Genomics* **4**, p. 44.

Liu, P. F., Jiang, W. H., Han, Y. T., He, L. F., Zhang, H. L., and Ren, H. (2015). Integrated microRNA-mRNA analysis of pancreatic ductal adenocarcinoma, *Genet. Mol. Res.* **14**, 3, pp. 10288–10297.

Ma, L., Huang, Y., Zhu, W., Zhou, S., Zhou, J., Zeng, F., Liu, X., Zhang, Y., and Yu, J. (2011). An integrated analysis of miRNA and mRNA expressions in non-small cell lung cancers, *PLoS ONE* **6**, 10, p. e26502.

Mahalingam, D., Kong, C. M., Lai, J., Tay, L. L., Yang, H., and Wang, X. (2012). Reversal of aberrant cancer methylome and transcriptome upon direct reprogramming of lung cancer cells, *Sci Rep* **2**, p. 592.

Meng, X. R., Lu, P., Mei, J. Z., Liu, G. J., and Fan, Q. X. (2014). Expression analysis of miRNA and target mRNAs in esophageal cancer, *Braz. J. Med. Biol. Res.* **47**, 9, pp. 811–817.

Miyanaga, A., Gemma, A., Noro, R., Kataoka, K., Matsuda, K., Nara, M., Okano, T., Seike, M., Yoshimura, A., Kawakami, A., Uesaka, H., Nakae, H., and Kudoh, S. (2008). Antitumor activity of histone deacetylase inhibitors in non-small cell lung cancer cells: development of a molecular predictive model, *Mol. Cancer Ther.* **7**, 7, pp. 1923–1930.

Moarii, M., Boeva, V., Vert, J.-P., and Reyal, F. (2015). Changes in correlation between promoter methylation and gene expression in cancer, *BMC Genomics* **16**, 1, pp. 1–14, doi:10.1186/s12864-015-1994-2, `http://dx.doi.org/10.1186/s12864-015-1994-2`.

Murakami, Y., Kubo, S., Tamori, A., Itami, S., Kawamura, E., Iwaisako, K., Ikeda, K., Kawada, N., Ochiya, T., and Taguchi, Y. H. (2015). Comprehensive analysis of transcriptome and metabolome analysis in Intrahepatic Cholangiocarcinoma and Hepatocellular Carcinoma, *Sci Rep* **5**, p. 16294.

Nagy, C. and Turecki, G. (2015). Transgenerational epigenetic inheritance: an open discussion, *Epigenomics* **7**, 5, pp. 781–790.

Rissman, E. F. and Adli, M. (2014). Minireview: transgenerational epigenetic inheritance: focus on endocrine disrupting compounds, *Endocrinology* **155**, 8, pp. 2770–2780.

Santos, A., Wernersson, R., and Jensen, L. J. (2015). Cyclebase 3.0: a multi-organism database on cell-cycle regulation and phenotypes, *Nucleic Acids Res.* **43**, Database issue, pp. D1140–1144.

Skinner, M. K., Guerrero-Bosagna, C., Haque, M., Nilsson, E., Bhandari, R., and McCarrey, J. R. (2013). Environmentally induced transgenerational epigenetic reprogramming of primordial germ cells and the subsequent germ line, *PLoS ONE* **8**, 7, p. e66318.

Taguchi, Y. H. (2015). Identification of aberrant gene expression associated with aberrant promoter methylation in primordial germ cells between E13 and E16 rat F3 generation vinclozolin lineage, *BMC Bioinformatics* **16 Suppl 18**, p. S16.

Taguchi, Y. H. (2016a). Identification of more feasible MicroRNA-mRNA interactions within multiple cancers using principal component analysis based unsupervised feature extraction, *Int J Mol Sci* **17**, 5, p. 696.

Taguchi, Y. H. (2016b). Principal component analysis based unsupervised feature extraction applied to budding yeast temporally periodic gene expression, *BioData Min* **9**, p. 22.

Taguchi, Y. H., Iwadate, M., and Umeyama, H. (2015a). Heuristic principal component analysis-based unsupervised feature extraction and its application to gene expression analysis of amyotrophic lateral sclerosis data sets, in *Computational Intelligence in Bioinformatics and Computational Biology (CIBCB), 2015 IEEE Conference on*, pp. 1–10, doi:10.1109/CIBCB.2015.7300274.

Taguchi, Y. H., Iwadate, M., and Umeyama, H. (2015b). Principal component analysis-based unsupervised feature extraction applied to in silico drug discovery for posttraumatic stress disorder-mediated heart disease, *BMC Bioinformatics* **16**, p. 139.

Taguchi, Y. H., Iwadate, M., and Umeyama, H. (2016). SFRP1 is a possible candidate for epigenetic therapy in non-small cell lung cancer, *BMC Medical Genomics* **9**, Suppl 1, p. 28, doi:10.1186/s12920-016-0196-3.

Taguchi, Y. H., Iwadate, M., Umeyama, H., Murakami, Y., and Okamoto, A. (2015c). Heuristic principal component analysis-aased unsupervised feature extraction and its application to bioinformatics, in B. Wang, R. Li, and W. Perrizo (eds.), *Big Data Analytics in Bioinformatics and Healthcare* (IGI Global, Pensylvania, USA), pp. 138–162.

Takahashi, A., Rahim, A., Takeuchi, M., Fukui, E., Yoshizawa, M., Mukai, K., Suematsu, M., Hasuwa, H., Okabe, M., and Matsumoto, H. (2016). Impaired female fertility in tubulointerstitial antigen-like 1-deficient mice, *J. Reprod. Dev.* **62**, 1, pp. 43–49.

Tang, Y. A., Wen, W. L., Chang, J. W., Wei, T. T., Tan, Y. H., Salunke, S., Chen, C. T., Chen, C. S., and Wang, Y. C. (2010). A novel histone deacetylase inhibitor exhibits antitumor activity via apoptosis induction, F-actin disruption and gene acetylation in lung cancer, *PLoS ONE* **5**, 9, p. e12417.

Tu, B. P., Kudlicki, A., Rowicka, M., and McKnight, S. L. (2005). Logic of the yeast metabolic cycle: temporal compartmentalization of cellular processes, *Science* **310**, 5751, pp. 1152–1158.

Umeyama, H., Iwadate, M., and Taguchi, Y.-h. (2014). TINAGL1 and B3GALNT1 are potential therapy target genes to suppress metastasis in non-small cell lung cancer, *BMC Genomics* **15**, Suppl 9, p. S2, doi:10.1186/1471-2164-15-S9-S2, `http://www.biomedcentral.com/1471-2164/15/S9/S2`.

Wang, A. and Gehan, E. A. (2005). Gene selection for microarray data analysis using principal component analysis, *Stat Med* **24**, 13, pp. 2069–2087.

Wu, B., Li, C., Zhang, P., Yao, Q., Wu, J., Han, J., Liao, L., Xu, Y., Lin, R., Xiao, D., Xu, L., Li, E., and Li, X. (2013). Dissection of miRNA-miRNA interaction in esophageal squamous cell carcinoma, *PLoS ONE* **8**, 9, p. e73191.

Yang, Y., Li, D., Yang, Y., and Jiang, G. (2015). An integrated analysis of the effects of microRNA and mRNA on esophageal squamous cell carcinoma, *Mol Med Rep* **12**, 1, pp. 945–952.

Zhang, W., Edwards, A., Fan, W., Flemington, E. K., and Zhang, K. (2012). miRNA-mRNA correlation-network modules in human prostate cancer and the differences between primary and metastatic tumor subtypes, *PLoS ONE* **7**, 6, p. e40130.

Chapter 9

Choquet integral algorithm for T-cell epitope prediction using support vector machine

Hsiang-Chuan Liu and Pei-Chun Chang*
Department of Bioinformatics and Medical Engineering, Asia University, Taichung 41354, Taiwan

9.1 Introduction

Antigen is a molecule that can induce immune response on the host organism to produce antibodies against it. Epitope are some segments in antigen which can replace the whole antigen to be recognized by the immune system specifically. The T-cell epitopes of protein antigens are some peptides that presented on the cell surface by MHC molecules via the MHC class I pathway or the MHC class II pathway in the immune mechanisms. The antigen presentation mechanisms are different in these two pathways. The MHC class I pathway is used to present endogenous antigens to cytotoxic T cell, while the MHC class II pathway is used to present the extracellular antigens to helper T cell. To accurately predict both types of epitopes is essential in vaccine design that is one goal of immunoinformatics.

Many methods for epitope prediction have been developed. These methods used the peptide features such as PSSM, quantitative matrix, entropy, pseudocount correction, or physicochemical properties, and combined with computational algorithm such as ANN [1], SVM [2], Gibbs sampling [3], or HMM [4]. In these methods, combining physicochemical properties of sequence and SVM-based methods have highest accuracy in T cell epitope prediction including MHC class I and class II. The accuracy of MHC-I binding predictors are currently in the

*corresponding author.

range of 90–95% positive prediction value. Nevertheless there is still limited success in predicting MHC-II-binding epitopes [5].

Fuzzy measure considers a series of special classes of measures and defined by a special property, respectively. These measures used in this theory may be plausibility and belief measures, fuzzy set membership function or the classical probability measures. In the fuzzy measure theory, the conditions are precise, but the information about an element alone is insufficient to determine which special classes of measure should be used. The concept of fuzzy measure theory was introduced by Choquet in 1953 and independently defined by Sugeno in 1974 in the context of fuzzy integrals [6]. The Choquet integral is a fuzzy integral based on any fuzzy measure that provides a computational scheme for information aggregation [7].

In this study, we supposed that the contribution of immunogenicity is determined by some different physicochemical properties for each amino acid in the peptide. These physicochemical properties in each amino acid may interact with each other and produce the total effect on immunogenicity. Based on this viewpoint, we proposed an algorithm combining fuzzy measure, Choquet integral, and SVM to predict the immunogenicity of peptides. Fuzzy measure and Choquet integral were used to estimate the total effect of immunogenicity regarding the interaction among these different physicochemical properties for each amino acid. The total effect of immunogenicity for each amino acid was defined as the element of the feature vector for the SVM classifier.

9.2 Methods

9.2.1 *Epitope datasets and physicochemical properties*

The dataset PEPMHCI of peptides [8] regarding to human MHC class I molecules was used. It was downloaded from POPI web server [9] and shows as Table 9.1. The immunogenicity of peptides belonging to the four classes, None, Little, Moderate, and High, we assigned the immunogenicity value with 0, 1, 2, and 3 for the four classes, respectively. The numbers of each peptide classes are 147, 95, 125, and 132, respectively. Since the interaction between epitope and MHC is conservative, the epitope sequences must be homogeneous. For this reason, the epitope sequences were aligned to determine its major binding positions. The sequence alignment processes were implemented in ClustalX 3.14 [10] for all the peptide sequences. For each position of the alignment sequence set, the amino acids were coded according to the Amino Acid Index database (AAindex) [11]. The alignment gaps were assigned to the average value of this position over all sequences. AAindex is a database of physicochemical and biochemical properties of amino acids derived from published literature. All data are represented as numerical indices to indicate the level of various physicochemical and biochemical properties. There are 544 physicochemical properties of amino acids extracted from amino acid index database (AAindex). The property having the value 'NA' in a value set of amino acid index was discarded. Finally, 531 properties were used in the coding process.

Table 9.1. The dataset PEPMHCI of peptides associated with human MHC class I molecules.

Immunogenicity class	Number of peptides
None	144
Little	83
Moderate	100
High	101
Total	428

9.2.2 *Fuzzy measure*

A fuzzy measure μ on a finite set X is a set function $g_\mu : 2^X \to [0,1]$ satisfying the following axioms:

- $g_\mu(\phi)=0,\ g_\mu(X)=1$ (boundary conditions) (1)
- $A \subseteq B \Rightarrow g_\mu(A) \le g_\mu(B)$ (monotonicity) (2)

g_μ is also called the measure function of μ.

The two well known fuzzy measures, the λ-measure proposed by Sugeno in 1974 [6], and L-measure proposed by Liu in 1978 [12], are concise introduced as follows

- λ-measure: For a given density function $s(x)$ on a finite set X, a λ-measure, g_λ, is a fuzzy measure on X, satisfying:

1) $\mu(\phi)=0, \mu(X)=1$ *(boundary conditions)*

2) $A,B \in 2^X, A \cap B = \phi, A \cup B \neq X$

$$\Rightarrow g_\lambda(A \cup B) = g_\lambda(A) + g_\lambda(B) + \lambda g_\lambda(A) g_\lambda(B) \quad (3)$$

3) $\prod_{i=1}^{n}[1+\lambda s(x_i)] = \lambda + 1 > 0,\ s(x_i) = g_\lambda(\{x_i\})$

Note that λ-measure has a unique solution without closed form. If $\sum_{x \in X} s(x) = 1$, then λ-measure is just additive measure.

- L-measure [14]: For any given density function $s(x)$ on a finite set X, $x \in X$, a L-measure, g_L, is a fuzzy measure on X, satisfying:

  1) $g_L(\phi) = 0, g_L(X) = 1$

  2) $L \in [0,\infty),\ \forall A \subset X, A \neq X \Rightarrow$

$$g_L(A) = \max_{x \in A}\{s(x)\} + \frac{L(|A|-1)\sum_{x \in A} s(x)\left[1 - \max_{x \in A}\{s(x)\}\right]}{\left[|X| - |A| + L(|A|-1)\right]\sum_{x \in X} s(x)} \quad (4)$$

In this study, we adopted λ-measure and L-measure, respectively.

9.2.3 *Choquet integral [14]*

Let μ be a fuzzy measure on a finite set X. The Choquet integral of $f_i : X \to R_+$ with respect to μ for individual i is denoted by

$$\int_C f_i d\mu = \sum_{j=1}^{n}\left[f_i\left(x_{(j)}\right) - f_i\left(x_{(j-1)}\right)\right]\mu\left(A_{(j)}\right), i = 1,2,\ldots,N \quad (5)$$

where $f_i\left(x_{(0)}\right) = 0$, $f_i\left(x_{(j)}\right)$ indicates that the indices have been permuted so that

$$0 \le f_i\left(x_{(1)}\right) \le f_i\left(x_{(2)}\right) \le \ldots \le f_i\left(x_{(n)}\right) \quad (6)$$

$$A_{(j)} = \left\{x_{(j)}, x_{(j+1)}, \ldots, x_{(n)}\right\} \quad (7)$$

9.2.4 *Feature extraction and support vector machine*

The feature window was defined as a segment of the aligned sequences which containing maximum variation of amino acids. We assessed the variation along the aligned sequences with window size of 9 amino acids by Shannon entropy (H) [15]:

$$H = -\sum_{n}\sum_{i} p_{ni} \log p_{ni} \tag{8}$$

where n represents the position in the window, and p_{ni} represents the proportion of i-th amino acid at position n.

For each position in the feature window of the aligned peptide sequences, normalize each physicochemical property for all peptide sequences at the same position. Assume the size of the peptide set is k. Let the i-th peptide sequence for physicochemical property m at position l be a variable $X_i^{l,m}$ where $1 \leqq l \leqq k$, $1 \leqq m \leqq 3$. If $\max_l\{X_i^{l,m}\} - \min_l\{X_i^{l,m}\} \neq 0$, then $Z_i^{l,m} = \dfrac{X_i^{l,m} - \min_l\{X_i^{l,m}\}}{\max_l\{X_i^{l,m}\} - \min_l\{X_i^{l,m}\}}$. Otherwise, set $Z_i^{l,m} = 0$.

Let Y_i be a variable which is the immunogenicity level for the i-th peptide sequence. Let $X^{l,m} = (X_1^{l,m}, X_2^{l,m}, \cdots, X_n^{l,m})'$ and $Y = (Y_1, Y_2, \cdots, Y_n)'$, where n is the total number of the aligned peptide sequences. For each m, compute corr($X^{l,m}$, Y) where "corr" is the Pearson's correlation coefficient. For each l, define the weight $w^{l,m}$ to be $\dfrac{1+corr(X^{l,m},Y)}{2}$ for each m. That is the fuzzy measure, $v(\{X^{l,m}\}) = w^{l,m}$ for $1 \leqq m \leqq 3$ and $1 \leqq l \leqq k$.

Finally, three physicochemical properties with highest Pearson's correlation of immunogenicity were extracted for each position in the feature window. These physicochemical properties of AAindex are shown as Table 9.2.

Table 9.2. The top three AAindex for each position in the feature window.

Position	AAindex ID	Description
–4	HUTJ700103	Entropy of formation
	HUTJ700102	Absolute entropy
	OOBM770102	Short and medium range non-bonded energy per atom
–3	TANS770106	Normalized frequency of chain reversal D
	QIAN880138	Weights for coil at the window position of 5
	AURR980116	Normalized positional residue frequency at helix termini Cc
–2	CHOP780215	Frequency of the 4th residue in turn
	WILM950104	Hydrophobicity coefficient in RP-HPLC, C18 with 0.1%TFA/2-PrOH/MeCN/H2O
	GEOR030102	Linker propensity from 1-linker dataset
–1	QIAN880124	Weights for beta-sheet at the window position of 4
	SNEP660101	Principal component I
	RACS820109	Average relative fractional occurrence in AL(i–1)
0	PRAM820101	Intercept in regression analysis
	GEOR030105	Linker propensity from small dataset (linker length is less than six residues)
	KRIW790102	Average relative fractional occurrence in AL(i–1)
+1	AURR980104	Normalized positional residue frequency at helix termini N'
	AURR980105	Normalized positional residue frequency at helix termini Nc
	QIAN880135	Weights for coil at the window position of 2
+2	RICJ880111	Relative preference value at C4
	CIDH920101	Normalized hydrophobicity scales for alpha-proteins
	PONP800106	Surrounding hydrophobicity in turn
+3	RICJ880101	Relative preference value at N"
	RICJ880102	Relative preference value at N'
	QIAN880102	Weights for alpha-helix at the window position of –5
+4	QIAN880138	Weights for coil at the window position of 5
	FINA910103	Helix termination parameter at position j–2,j–1,j
	CHAM830107	A parameter of charge transfer capability

To combine the three physicochemical properties to be one for each amino acid in an epitope sequence, we compute the Choquet integral for each position by Equation (5). Then we get one feature vector for each aligned epitope sequence.

Support vector machine (SVM) is a learning model dealing with classification problems. We adopted SVM model to predict the immunogenicity class according to the feature vector of epitope sequence. In order to make linear separation of samples easier, SVM constructs a classifier by finding a hyperplane to separate two or multiple classes after mapping the sampling points into high-dimensional space. The mapping process can be defined by various kernel functions. In our study, the radial basis function was adopted to nonlinearly transform the feature space:

$$K(x_i, x_j) = \exp(-\gamma \| x_i - x_j \|), \ \gamma > 0 \tag{9}$$

The kernel parameter determines how the samples are transformed into a high-dimensional sampling space. The cost parameter C>0 of SVM adjusts the total error penalty. The parameters C and γ must be tuned to get the best prediction performance [16].

9.3 Results

For evaluating the performance of this new algorithm, the PEPMHCI dataset by using 5-fold Cross-Validation was conducted. Three measurements were used, namely percentage accuracy (ACC_i), for the *i*-th immunogenicity class, *i*=1, . . . , 4, overall accuracy (OA) and averaged accuracies (AA) for all classes:

$$ACC_i = \frac{TP_i}{TP_i + FN_i} \tag{10}$$

$$OA = \sum \frac{TP_i}{N} \tag{11}$$

$$AA = \sum \frac{ACC_i}{h} \tag{12}$$

where TP_i, TN_i, FP_i and FN_i are the number of true positive, true negative, false positive and false negative, respectively. N is the total number of peptide sequences and h is the number of immunogenicity classes.

Table 9.3 shows the performance of our algorithm in term of ACC for the four immunogenicity classes, and the prediction accuracies of OA and AA. In the case of Lambda measure, the ACC accuracies of the four classes None, Little, Moderate and High are 94.74, 89.47, 74.00 and 96.77%, respectively. The overall accuracy and average accuracy are 92.05 and 88.79%, respectively. The other case of L-measure, the ACC accuracies of the four classes None, Little, Moderate and High are 94.73, 81.57, 72.00 and 95.167%, respectively. The overall accuracy and average accuracy are 90.18 and 85.86%, respectively.

As the results, our prediction methods based on Lambda measure and L-measure (L=0.6) have better performance than POPI [9] for every immunogenicity classes.

Table 9.3. Performance comparisons of POPI our algorithm with lambda measure, and our algorithm with l measure using 5-fold cross-validation on the whole dataset PEPMHCI.

Immunogenicity class	POPI	Lambda measure	L-measure (L=0.6)
	ACC (%)	ACC (%)	ACC (%)
None	83.33	94.74	94.73
Little	50.60	89.47	81.57
Moderate	55.00	74.00	72.00
Hight	59.41	96.77	95.16
AA	64.72	88.79	85.86
OA	62.09	92.05	90.18

9.4 Discussion

The peptide immunogenicity in peptide-based vaccine design is important in vaccine development. Accurate prediction of peptide immunogenicity will decrease the cost in the experimental screens.

This study investigates the prediction problem of peptide immunogenicity based on the assumption of coupling effects among physicochemical properties in an amino acid and proposes an efficient prediction algorithm to predict immunogenicity of peptides with variable lengths. The coupling effects were assessed by fuzzy measure and Choquet integral. In the last step, the SVM classifier was used in the prediction method. A dataset PEPMHCI of peptides associated with human MHC class I molecules extracted from MHCPEP was evaluated. The results in this study implied that the coupling effects among physicochemical properties in an amino acid were important in peptide immunogenicity.

References

1. Kemir, C., Nussbaum, A. K., Schild, H., Detours, V., and Brunak, S. (2002) Prediction of proteasome cleavage motifs by neural networks, *Protein Eng.*, 15, pp. 287–296.
2. Bhasin M. and Raghava, G. P. (2005) Pcleavage: an SVM based method for prediction of constitutive proteasome and immunoproteasome cleavage sites in antigenic sequences, *Nucleic Acids Res.*, 33, pp. W202–W207.
3. Nielsen, M., Lundegaard C., Worning, P., Hvid, C. S., Lamberth, K., Buus, S., Brunak, S., and Lund, O. (2004) Improved prediction of MHC class I and class II epitopes using a novel Gibbs sampling approach, *Bioinformatics*, 20, pp. 1388–1397.
4. Larsen, M. V., Lundegaard, C., Lamberth, K., Buus, S., Brunak, S., Lund, O., and Nielsen, M. (2005) An integrative approach to CTL epitope prediction: a combined algorithm integrating MHC class I binding, TAP transport efficiency, and proteasomal cleavage predictions, *Eur. J. Immunol.*, 35, pp. 2295–2303.
5. Lin, H. H., Zhang, G. L., Tongchusak, S., Reinherz, E. L., and Brusic. V. (2008) Evaluation of MHC-II peptide binding prediction servers: applications for vaccine research, *BMC Bioinformatics*, 9, pp. S22.

6. Sugeno, M. (1974) Theory of fuzzy integrals and its applications, *unpublished doctoral dissertation*. (Tokyo Institute of Technology, Tokyo, Japan).
7. Murofushi T. and Sugeno, M. (1993) Some quantities represented by the choquet integral, *Fuzzy Sets Syst.*, 56, pp. 229–235.
8. Brusic, V., Rudy, G., and Harrison, L. C. (1998) MHCPEP, a database of MHC-binding eptides:update 1997, *Nuceic Acids Research*, 25, pp. 269–271.
9. Tung, C. W. and Ho, S. Y. (2007) POPI:predicting immunogenicity of MHC class I binding peptides by mining informative physicochemical properties, *bioinformatics*, 23, pp. 942–949.
10. Thompson, J. D., Gibson, T. J., and Higgins, D. G. (2002) Multiple sequence alignment using ClustalW and ClustalX, *Curr. Protoc. Bioinformatics*, Chapter 2.
11. Shuichi, K., Piotr, P., Maria, P., Andrzej, K., Toshiaki, K., and Minoru, K. (2008) AAindex:amino acid index database, progress report 2008, *Nuceic Acids Research*, 36, pp. D202–D205.
12. Zadeh, L. A. (1978) Fuzzy sets as a basis for a theory of possibility, *Fuzzy Sets and Systems*, 1, pp. 3–28.
13. Liu, H. C., Tu, Y. C., Lin, W. C., and Chen, C. C. (2008) Choquet integral regression model based on L-Measure and γ-Support, *Proceedings of 2008 International Conference on Wavelet Analysis and Pattern Recognition*, ICWAPR, pp. 777–782 (in Hong Kong).
14. Choquet, G. (1953) Theory of capacities, *Annales de l'Institut Fourier*, 5, pp. 131–295.
15. Shannon, C. E. (1951) Prediction and entropy of printed English, *The Bell System Technical Journal*, 30, pp. 50–64.
16. Fan, R. E., Chen, P. H., and Lin, C. J. (2005) Working set selection using second order information for training SVM, *Journal of Machine Learning Research*, 6, pp. 1889–1918.

Chapter 10

Unsupervised clustering algorithms for flow/mass cytometry data

Jinmiao Chen[1,*] and Feng Lin[2]

[1]*Singapore Immunology Network, Agency for Science, Technology and Research (A*STAR), Singapore*

[2]*School of Computer Engineering, Nanyang Technological University, Singapore*

[*]*chen_jinmiao@immunol.a-star.edu.sg*

In this chapter, we introduce the latest high-dimensional flow and mass cytometry technologies, and review state-of-the-art unsupervised clustering algorithms for this type of data. We continue to evaluate and compare three most recent algorithms on a mass cytometry data set, and discuss challenges that remain.

10.1 Introduction

Flow and mass cytometry are widely used in clinical and basic research to characterize cell phenotypes and functions. Both measure the expression of surface and intracellular molecules (termed "markers") in individual cells. Flow cytometry is a laser-based cytometric technique in which cells are stained with fluorescence-conjugated antibodies and taken past a laser light one cell at a time by a tiny stream of fluid. As the cell is passing through the laser beam, the cell will scatter the light; the fluorochromes will emit light when excited by a laser with the corresponding excitation wavelength. The intensity of scattered and fluorescent light is detected and analysed. Mass cytometry (a.k.a. CyTOF) is a next-generation flow cytometer that uses heavy metal isotopes to tag antibodies instead of fluorophores. By using isotopic tagging, mass cytometry produces little crosstalk between channels as compared to flow cytometry.

The main aim of flow/mass cytometry data analysis is to identify cell subsets based on marker expression of individual cells. The conventional method of flow and mass cytometry data analysis is manual gating using software such as FCS Express, FlowJo, FACSDiva and etc. Using the manual gating software, scientists examine biaxial plots of every two markers and manually created a series of cell subset extractions based on fluorescence intensity. In theory, $m(m-1)/2$ biaxial plots are needed to be examined for data of m markers to fully define all cell populations. Recent development of cytometry technologies has significantly increased the number of markers per cell. For instance, most current flow cytometers can measure 16-18 markers per cell; the latest BD FACSSymphony is able to analyze 40+ markers per cell; mass cytometry offers even more available channels (>40 markers per cell). The emerging high-dimensional flow and mass cytometry produces high-dimensional data, which poses a significant technical challenge for data analysis as manual gating of over 40 markers becomes prohibitively impractical.

Another caveat of manual gating lies in its lack of objectivity and reproducibility. It could be biased by prior knowledge and preference of the individual. Manual gating of the same data can vary between different individuals who perform the analysis. Even the same individual can also produce different analysis on different days. Moreover, by manual gating, we usually identify known cell types based on known defining markers, but fail to discover novel cell types. On the other hand, manual gating is a time consuming and labour intensive process. More and more large-scale flow/mass cytometry experiments analyse hundreds or thousands of samples. Manual gating of large sample sets becomes tedious and inefficient. Objective and automated methods other than manual gating are thus highly desired.

To overcome the limitations of manual gating, automated, objective and unsupervised methods of cell subset identification are highly in demand. Unsupervised clustering algorithms allow automated grouping of cells into subsets based on similar expression of markers. A number of clustering algorithms have been designed specifically for flow and later mass cytometry data. These algorithms fall into two major categories, with and without dimensionality reduction. A number of dimensionality reduction methods have been proposed including PCA, ISOMAP, t-SNE,[1] diffusion map and etc. Among them, t-SNE has been demonstrated to be most effective in segregating cell populations in the reduced dimension.

For flow cytometry data, the very first algorithm called flowClust,[2] a Bioconductor package for automated gating of flow cytometry data was proposed. flowClust implements a robust model-based clustering approach based on multivariate *t* mixture models with the Box-Cox transformation. By using multivariate *t* mixture models instead of the most commonly used finite Gaussian mixture models, flowClust is able to identify outliers as well as clusters that are far from elliptical shape. One key challenge of mixture model based clustering is to determine the optimal number of clusters. The max BIC model fitting criterion generally overestimates the number of clusters; whilst model fitting criteria based on the entropy, such as the ICL, tend to provide poor fit to the underlying distribution. Thus a Bioconductor package called flowMerge combines these two approaches to achieve good model fitting and accurately estimate number of clusters. flowMerge first chooses the best BIC solution, then merges clusters in the best BIC solution, and choose the best merged solution based on the entropy criterion. On the other hand, Model-independent or non-parametric clustering method, such as spectral clustering, has the advantage in not requiring a priori assumption that cell populations follow the predefined distributional models. However, spectral clustering is computationally intensive and time inefficient for large datasets. In order to improve efficiency, SamSPECTRAL[3] modified spectral clustering by a non-uniform information preserving down-sampling. Another model-independent approach, FLOw Clustering without K (FLOCK),[4] utilizes grid-based partitioning and density distribution analysis to identify cell populations. It partitions the n-dimensional space into "hyperregions" by partitioning each dimension into equally sized bins. Any hyperregion in which the cell count exceeds a pre-defined threshold is labeled as ''dense'' hyperregion. Adjacent "dense" hyperregions are then merged. Each cell is then assigned to the nearest centroids of the merged "dense" hyperregions. K-means is widely used for clustering. However it requires a predefined number of K. flowMeans[5], based on K-means, first uses kernel density based mode detection and uses the number of modes as K to run K-means clustering. The number of modes usually overestimates the number of clusters. flowMeans iteratively merges the closest pair of clusters based on a symmetric Mahalanobis semi-metric distance between clusters, until the distance between the next clusters to be merged is significantly larger than the previous one. This change point in the distance between the merged clusters is detected using a segmented regression algorithm. flowPeaks[6] also uses K-means with a large K to generate many clusters. Based on the number of cells in each cluster, it uses finite mixture model to approximate a smoothed density function. Subsequently it uses the greatest gradient search (hill climbing) to find local density peaks. The cells are then clustered by the associated local peaks. Scalability to large-sized datasets and identification of rare cell populations are the two common challenges for automated clustering. Scalable Weighted Iterative Flow-clustering Technique (SWIFT)[7] addresses these challenges by incorporating model-based clustering with iterative weighted sampling, multimodality splitting and unimodality-preserving merging. One more big challenge for analysis of these data is the matching of identified clusters between samples. One potential solution is to pool cell events from all samples and then run clustering. However this pooling strategy has disadvantages when we have thousands of samples and batch variation. To address this challenge, immunoClust[8] proposed to perform two clustering steps, first to cluster cell events in a sample followed by meta clustering of the cell-clusters from different samples.

With the emerging of mass cytometry that produces data of higher dimension, a number of novel algorithms have been developed specially for analysing this type of data. Spanning-tree progression analysis of density-normalized events (SPADE)[9] first performs density-dependent downsampling and then agglomerative clustering to partition the downsampled cells into clusters, and lastly performs minimum spanning tree to construct the geometry of the clusters. Automatic classification of cellular expression by nonlinear stochastic embedding (ACCENSE)[10] performs tSNE dimension reduction followed by density-based clustering on the 2D tSNE map. ACCENCE used peak-finding algorithm to detect local density peaks which become the cluster centroid. Cells whose distance to peak k is less than half of the distance between peak k and its nearest neighbor peak are assigned to cluster k. By ACCENSE, a significant number of cells are not assigned to any clusters. On the basis of ACCENSE's cluster analysis, DensVM grouped cells into training set and testing set. The training set contained cells that received cluster assignment from ACCENSE; cells that ACCENSE failed to classify comprised the testing set. DensVM[11] used support vector machine (SVM) to train a classifier that learns the mapping from marker expression profile to cluster assignment. Under the assumption that cells with similar marker expression originated from the same cluster, the trained classifier took as input marker expression values of cells in the testing set and generated cluster predictions for these cells. DensVM obtained the final cluster delineation by combing the cluster assignment given by ACCENSE with the cluster prediction made by support vector machine. By this method, cells sharing similar patterns of marker expression are classified into the same cluster (or subpopulations). By incorporating machine learning–aided clustering, DensVM is able to precisely detect the boundaries of cell populations and hence allows their frequencies to be objectively compared. Both DensVM and the original form of ACCENSE required an iterative search for the optimal bandwidth that was used to estimate kernel density. Another similar method named ClusterX[12] is also density-based clustering on tSNE project map. By applying Clustering by fast search and find of density peaks (CFSFDP) algorithm, ClusterX doesn't require iterative search for the optimal bandwidth for density estimation, and hence improves time efficiency over ACCENSE and DensVM. An even faster clustering method named FlowSOM[13] first performs self-organizing map clustering and subsequently merges clusters by hierarchical consensus meta-clustering. Similar to K-means, FlowSOM requires users to specify the number of clusters. A recent method named phonograph[14] has been demonstrated to be effective on multiple datasets. It first transforms the high-dimensional data into a nearest-neighbour graph in which each node represent one cell and cells are connected by edges to a neighborhood of its most similar cells. Adopting the idea of partitioning large social networks into communities, phenograph clusters the cells by partitioning the nearest-neighbour graph into sets of highly interconnected nodes. Many of these methods for mass cytometry involve a step of dimension reduction. For instance, ACCENSE, DensVM and ClusterX are based on tSNE dimension reduction, while phenograph performs dimension reduction to a nearest-neighbor graph. One very recent method called X-shift[15] uses fast k-nearest-neighbor estimation of cell event density; then searches for local density maxima which become cluster centroids; lastly connects all the remaining cells to the centroids via density-ascending paths.

In the following sections, we performed validation and comparison of three clustering algorithms including phonograph, clusterX and flowSOM on a CyTOF data set from mouse bone marrow.

10.2 Data description

Ten replicate bone marrow samples were harvested from C57BL/6J mice, stained with a panel of 39 antibodies to cell surface markers, profiled by CyTOF and manually gated by cytometry experts to identify 24 immune cell populations.[15] FCS files and gates were downloaded from http://web.stanford.edu/~samusik/vortex/.

10.3 Method

Raw intensity values were subject to noise thresholding and asinh transformation $y = \text{asinh}(\frac{\max(x-1,0)}{5})$. 20,000 cell events per FCS file were randomly sampled and concatenated. The concatenated cell events were subject to t-SNE dimensionality reduction using all 39 surface markers. ClusterX cluster analysis was performed on t-SNE reduced dimension 1 and 2. Phenograph and flowSOM cluster analysis were performed on the original dimension using all 39 surface markers. ClusterX and Phenograph automatically determined the optimal number of clusters. FlowSOM was run with number of clusters k set to 25.

Given the set of clusters and hand-gated cell populations, contingency matrix C was computed, where C_{ij} is the number of cells in the i-th cluster that belong to j-th population. F-measure matrix $F_{ij} = 2(R_{ij}P_{ij})/(R_{ij} + P_{ij})$ was computed, where $P_{ij} = C_{ij}/\sum_k C_{ik}$ is the precision matrix and $R_{ij} = C_{ij}/\sum_k C_{kj}$ is the recall matrix. An optimal one-to-one assignment between clusters and gated populations was determined by running the Hungarian algorithm on the negative matrix $F' = 1 - F$, so that the sum of F-measures was maximized. The codes of Hungarian algorithm were obtained from http://software-and-algorithms.blogspot.com.

10.4 Results

t-SNE offers a way to visualize the underlying cell populations. Applied to the CyTOF data set, t-SNE reduced the original 39 dimensions to 2 dimensions; on which similar cells were placed nearby and dissimilar cells far part (Figure 1). We overlaid hand-gated cell populations on the t-SNE map (Figure 2). Each cell population occupied distinct regions on the t-SNE plot, indicating the t-SNE is able to segregate known populations. However, some populations such as pro-B cells and eosinophils scatter across more than one region. This could be due to the limitation of t-SNE or the fact that these cells are heterogonous and comprise sub-populations.

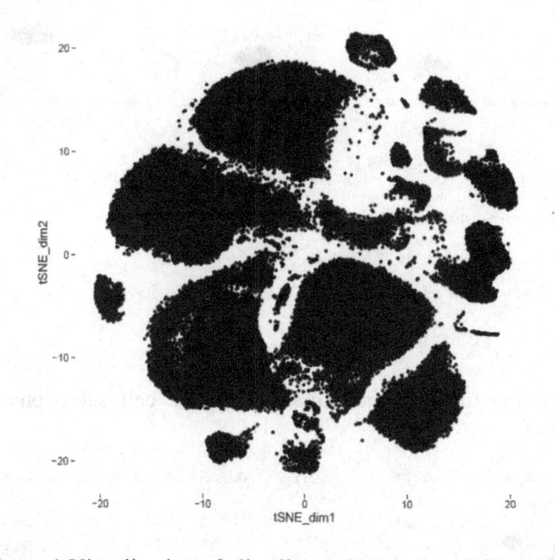

Figure 1 Visualization of all cells on tSNE-reduced dimensions

We continued to perform unsupervised clustering to group cells into clusters. We first applied the ClusterX algorithm which identifies density peaks on t-SNE plot and assigns individual cells to the nearest density peak (Figure 3). ClusterX clusters are highly concordant to t-SNE plot. It was able to partition t-SNE plot into distinct regions and detect the boundaries precisely. We ran flowSOM clustering using all the 39 markers without dimension reduction. Most flowSOM clusters are positioned within one region of the t-SNE map (Figure 4), except several clusters such as 13, 18, 25 were positioned in multiple regions. Similarly, phenograph clustering using all markers also produced clusters that can be visually verified by t-SNE plot (Figure 5).

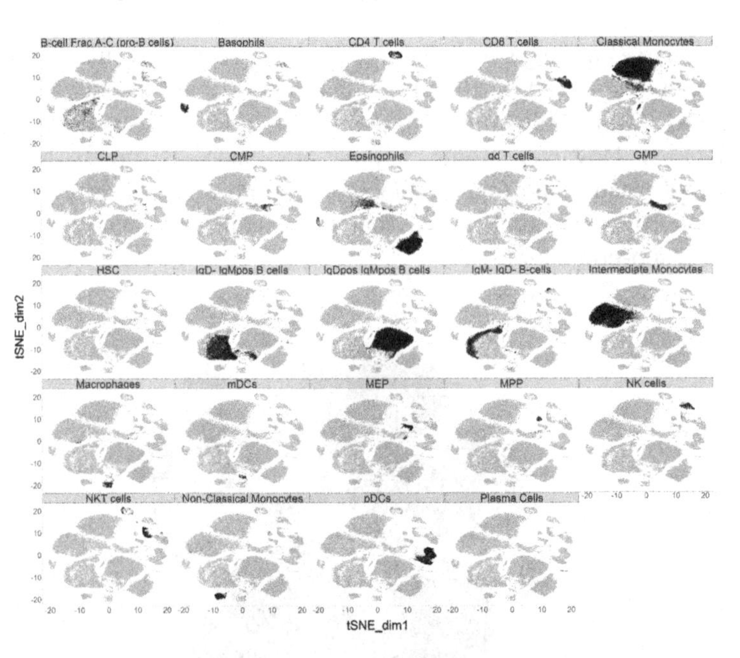

Figure 2 Visualization of cells from manually gated cell sub-population on t-SNE reduced dimensions

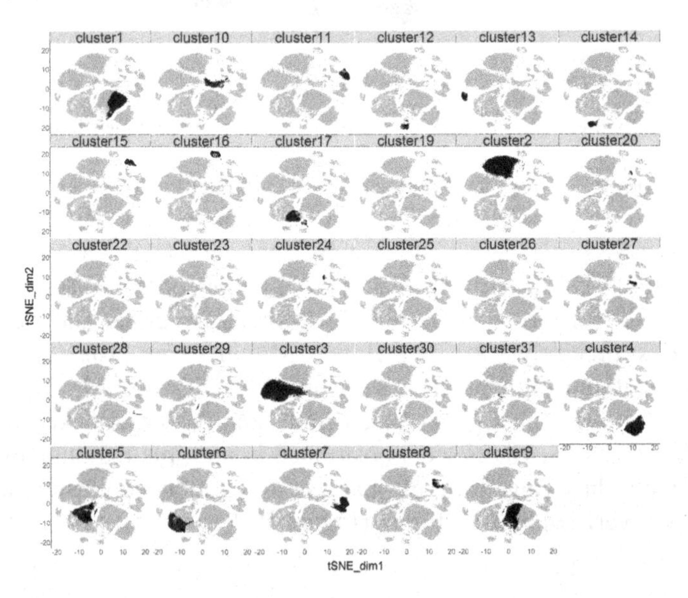

Figure 3 Visualization of cells from ClusterX algorithm determined clusters on t-SNE reduced dimensions

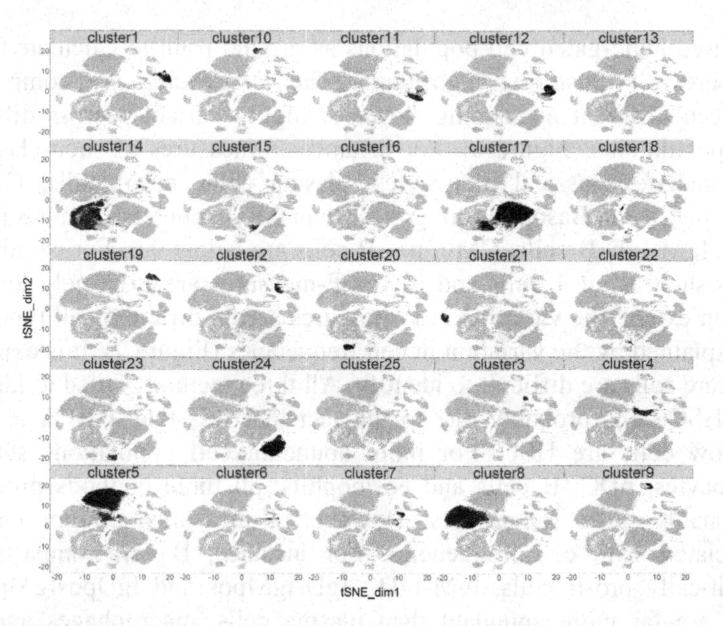

Figure 4 Visualization of cells from FlowSOM algorithm determined clusters on t-SNE reduced dimensions

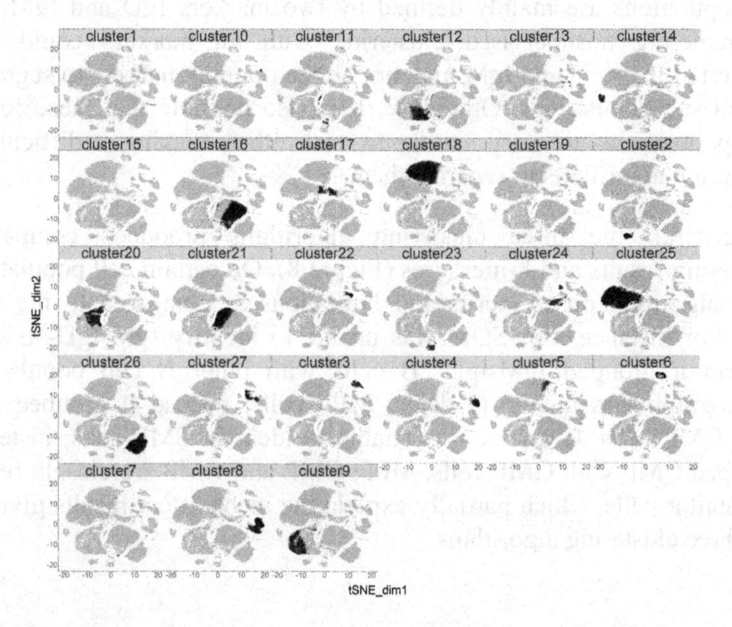

Figure 5 Visualization of cells from phenograph algorithm determined clusters on t-SNE reduced dimensions

We used hand-gated cell populations as ground truth to calculate the F-measures of clustering algorithms. The F-measures are comparable between different algorithms, however highly variable across different cell populations (Figure 6). For instance, F-measures of non-classical, intermediate, classical monocytes, plasma cells, eosinophils, CD8 T cells, pDCs and Basophils are greater than 0.9. In contrast, F-measures of CLP, IgM-IgD-B cells, CMP, pro-B cells are below 0.6. For certain cell types such as gd T cells and HSCs, F-measures are extremely low. To certain extent, the variation in F-measures across different cell types can be explained by the variation in cell frequencies (Figure 7). It is expected that rare cells are difficult to identify. All three methods failed to identify any HSC cells probably due to the fact that less than 0.01% of bone marrow cells are HSC. For more abundant cell populations such as monocytes, pDC, B cells and eosinophils, all three methods produced reasonably good F-measures. However, F-measures are not entirely associated with cell frequencies. For instance, B cell compartments specifically pro-B cells, IgM-IgD-, IgD-IgMpos and IgDposIgMpos B cells are far more abundant than plasma cells, macrophages and NK cells, but have lower F-measures. The manual gates of B cell subpopulations are mainly defined by two markers IgD and IgM. The contradicting unsupervised clustering with all markers could have suggested that IgD and IgM may not be the optimal markers to segregate B cell sub-populations. Otherwise, it is also possible that these four B cell populations are very similar to each other, making it difficult for clustering algorithms to separate them.

In general, the three clustering algorithms produced comparable clustering results and F-measures (Figure 8). On certain cell populations, each algorithm poses merits and limitations as compared to the other two. For instance, flowSOM was unable to identify IgM-IgD- B cells; instead it grouped IgM-IgD- B cells with other B cell populations. Phenograph was unable to detect MPP cells; instead it grouped MPP with CMP cells. ClusterX was unable to identify CMP cells; instead it grouped CMP with GMP cells. MPP, CMP and GMP are closely related progenitor cells, which partially explain the inconsistent results given by the three clustering algorithms.

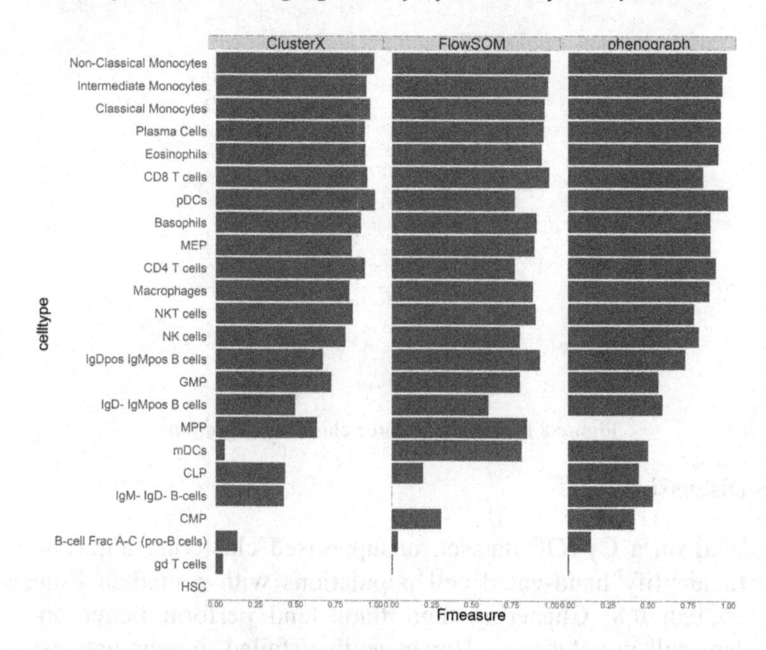

Figure 6 F measure of clustering algorithms for individual cell populations

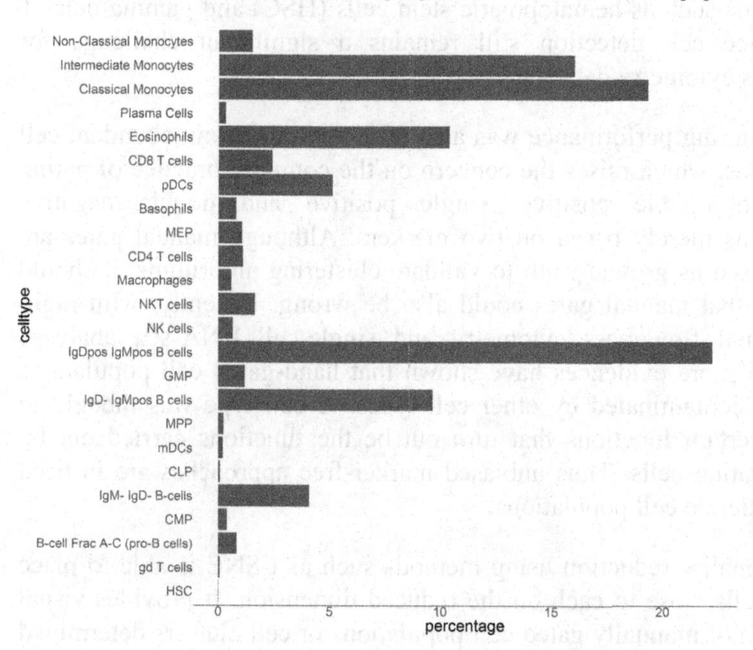

Figure 7 Percentage of hand-gated cell populations

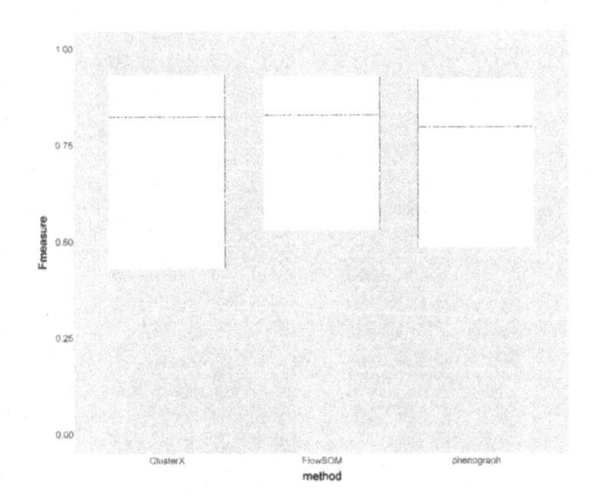

Figure 8 F-measure of three clustering algorithms

10.5 Discussion

Validated on a CyTOF dataset, unsupervised clustering algorithms are able to identify hand-gated cell populations with a median F-measure greater than 0.8. Clustering algorithms tend perform better on more abundant cell populations. However, they failed to segregate rare cell populations such as hematopoietic stem cells (HSC) and gamma delta T cells. Race cell detection still remains a significant challenge for flow/mass cytometry data analysis.

Poor clustering performance was also observed on certain abundant cell populations, which raises the concern on the common practice of gating cells into double positive, single positive and double negative populations merely based on two markers. Although manual gates are usually used as ground truth to validate clustering algorithms, it should be noted that manual gates could also be wrong. Recently, with high-dimensional flow/mass cytometry and single-cell RNA-seq analysis, more and more evidences have shown that hand-gated cell populations could be contaminated by other cell types. A cell type was thought to present certain functions that turn out be the functions carried out by contaminating cells. Thus unbiased marker-free approaches are in need to better define cell populations.

Dimensionality reduction using methods such as t-SNE is able to place similar cells close to each on the reduced dimension. It provides visual inspection of manually gated cell populations or cell clusters determined from unsupervised clustering. As ClusterX algorithm is based on t-SNE dimensions, its clusters are usually consistent with t-SNE map. However, it is not the case for algorithms such as phenograph and flowSOM that cluster cells based on the original dimensions. It is not unusual that phenograph or flowSOM clusters don't align well with t-SNE map. For some datasets, t-SNE and t-SNE based clustering perform better; while for some other datasets, t-SNE independent clustering performs better. The performance of dimension reduction based methods such as ACCENSE, DensVM and clusterX, will be affected by the performance of dimension reduction. Both dimension reduction and clustering algorithms need further improvement. New algorithms need to be developed to perform optimal dimension reduction and clustering at the same time.

References

1 van der Maaten, L. & Hinton, G. Visualizing High-Dimensional Data Using t-SNE. *Journal of Machine Learning Research* **9**, 2579-2605 (2008).

2 Lo, K., Hahne, F., Brinkman, R. R. & Gottardo, R. flowClust: a Bioconductor package for automated gating of flow cytometry data. *BMC bioinformatics* **10**, 145, doi:10.1186/1471-2105-10-145 (2009).

3 Zare, H., Shooshtari, P., Gupta, A. & Brinkman, R. R. Data reduction for spectral clustering to analyze high throughput flow cytometry data. *BMC bioinformatics* **11**, 403, doi:10.1186/1471-2105-11-403 (2010).

4 Qian, Y. *et al.* Elucidation of seventeen human peripheral blood B-cell subsets and quantification of the tetanus response using a density-based method for the automated identification of cell populations in multidimensional flow cytometry data. *Cytometry. Part B, Clinical cytometry* **78 Suppl 1**, S69-82, doi:10.1002/cyto.b.20554 (2010).

5 Aghaeepour, N., Nikolic, R., Hoos, H. H. & Brinkman, R. R. Rapid cell population identification in flow cytometry data. *Cytometry. Part A : the journal of the International Society for Analytical Cytology* **79**, 6-13, doi:10.1002/cyto.a.21007 (2011).

6 Ge, Y. & Sealfon, S. C. flowPeaks: a fast unsupervised clustering for flow cytometry data via K-means and density peak finding. *Bioinformatics* **28**, 2052-2058, doi:10.1093/bioinformatics/bts300 (2012).

7 Naim, I. *et al.* SWIFT-scalable clustering for automated identification of rare cell populations in large, high-dimensional flow cytometry datasets, part 1: algorithm design. *Cytometry. Part A : the journal of the International Society for Analytical Cytology* **85**, 408-421, doi:10.1002/cyto.a.22446 (2014).

8 Sorensen, T., Baumgart, S., Durek, P., Grutzkau, A. & Haupl, T. ImmunoClust — An automated analysis pipeline for the identification of immunophenotypic signatures in high-dimensional cytometric datasets. *Cytometry. Part A : the journal of the International Society for Analytical Cytology* **87**, 603-615, doi:10.1002/cyto.a.22626 (2015).

9 Qiu, P. *et al.* Extracting a cellular hierarchy from high-dimensional cytometry data with SPADE. *Nature biotechnology* **29**, 886-891, doi:10.1038/nbt.1991 (2011).

10 Shekhar, K., Brodin, P., Davis, M. M. & Chakraborty, A. K. Automatic Classification of Cellular Expression by Nonlinear Stochastic Embedding (ACCENSE). *Proceedings of the National Academy of Sciences of the United States of America* **111**, 202-207, doi:10.1073/pnas.1321405111 (2014).

11 Becher, B. *et al.* High-dimensional analysis of the murine myeloid cell system. *Nature immunology* **15**, 1181-1189, doi:10.1038/ni.3006 (2014).

12 Chen, H. *et al.* Cytofkit: A Bioconductor Package for an Integrated Mass Cytometry Data Analysis Pipeline. *PLoS computational biology* **12**, e1005112, doi:10.1371/journal.pcbi.1005112 (2016).

13 Van Gassen, S. *et al.* FlowSOM: Using self-organizing maps for visualization and interpretation of cytometry data. *Cytometry. Part A : the journal of the International Society for Analytical Cytology* **87**, 636-645, doi:10.1002/cyto.a.22625 (2015).

14 Levine, J. H. *et al.* Data-Driven Phenotypic Dissection of AML Reveals Progenitor-like Cells that Correlate with Prognosis. *Cell* **162**, 184-197, doi:10.1016/j.cell.2015.05.047 (2015).

15 Samusik, N., Good, Z., Spitzer, M. H., Davis, K. L. & Nolan, G. P. Automated mapping of phenotype space with single-cell data. *Nature methods* **13**, 493-496, doi:10.1038/nmeth.3863 (2016).

Index

V

W

X

Y

Z

Printed in the United States
By Bookmasters